Mathematical Measures
Mathematical Pleasures
Mathematical Treasures

Second Edition

By Stephen C. Clark

"Geometry has two great treasures; one is the theorem of Pythagoras; the other, the division of a line into extreme and mean ratio. The first we may compare to a measure of gold, the second we may name a precious jewel." Johannes Kepler

The Triangle's Sides

3	4	5
Skipped Fibonacci	No.	8
5	12	13
Skipped Fibonacci	No.	21
16	30	34
Skipped Fibonacci	No.	55
39	80	89
Skipped Fibonacci	No.	144
105	208	233
Skipped Fibonacci	No.	377
272	546	610
Skipped Fibonacci	No.	987
715	1428	1597
Skipped Fibonacci	No.	2584
1869	3740	4181

The cover illustration, our symbol since 2006, has both of Kepler's mathematical treasures. We used the 3,4,5 Triangle to demonstrate the Pythagorean Theorem and the 1, 2, $\sqrt{5}$ Triangle for the Golden (Extreme and Mean) Ratio. We didn't know at the time that the two triangles are related. The ratios of the sides of the 1,2,$\sqrt{5}$ Triangle are the culmination of a sequence that begins with the 3,4,5 Triangle. The series is a variation of the Fibonacci Sequence (See section IV).

Side/Column 1 = The Difference between the Skipped Fibonacci Number and the Previous Number in the First Column

8 - 3 = 5   21 - 5 = 16

55 - 16 = 39   144 - 39 = 105

Side 2 = The sum of the 3 Previous Sides

3 + 4 + 5 = 12   5 + 12 + 13 = 30

34 + 30 + 16 = 80   89 + 80 + 39 = 208

Side 3 = Every other Fibonacci Number beginning with 5, 8, 13, 21, 34, 55, 89, 144, 233, 377, 610, 987, 1597, 2584, 4181

The Ratio of the Sides in the 1, 2, $\sqrt{5}$ Triangle is

1/1 = 1   2/1 = 2 $\sqrt{5}$/1 = 2.236..

The Ratios of the Sides in the Sequence

3/3 = 1   4/3 = 1.333..   5/3 = 1.666..

5/5 = 1   12/5 = 2.4   13/5 = 2.6

16/16 =1   30/16 =1.875   34/16 = 2.125

39/39 = 1   80/39 = 2.051..   89/39 = 2.282..

105/105 = 1   208/105 = 1.981..   233/105 = 2.219

272/272 = 1   546/272 = 2.007..   610/272 = 2.243..

715/715 = 1   1428/715 = 1.997..   1597/715 = 2.234..

1869/1869 = 1   3740/1869 = 2.001..   4181/1869 = 2.237..

What this process has taught us is that everything in math is related

Table of Contents

 Math for the Ages ISBN # 9780615743752

Introduction to the Second Edition

The Second Edition covers much of the same territory as the first edition but with the added perspective that using the first edition has given us.

<u>The Circle:</u> It still remains the center of our efforts, but we introduce a new tool, the Circle Maker/Focus Stick.

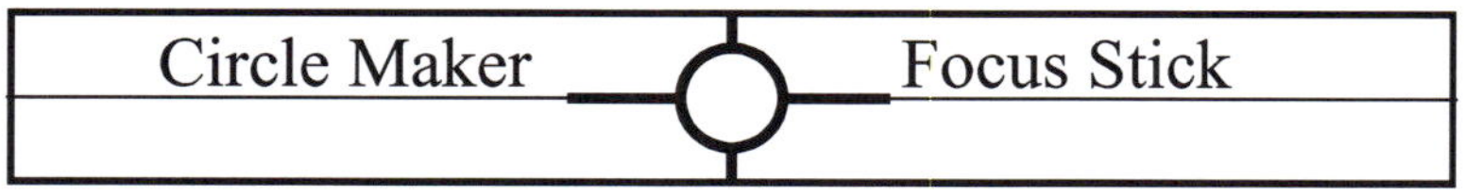

We use this tool instead of a Compass because it gives the learner a more direct experience of constructing the circle and provides insights into the nature of the circle that are simply not available through other means. We use our constructed circle to derive the Regular Polygons. The Regular Polygons are used to create the exterior faces of the Regular Polyhedra, also known as the Platonic Solids. We move through one, two, and three dimensions with simple projects that involve drawing, cutting, and folding, to produce an integrated mathematical universe.

<u>The Square:</u> While the circle/sphere is the building block of nature, the Square is the basic building block of human conceptualization. We use the simple form, the unit square, 1x1, to derive all of our shapes in this section. The Unit Square gives us the next four regular squares, 2x2, 3x3, 4x4, and 5x5 and the 3,4,5 Triangle and its companion the Pythagorean Theorem. We divide the unit square along its diagonal which yields the $1,1,\sqrt{2}$ triangle, the first member in the Spiral of Theodorus. Our other three triangles $1,\sqrt{2},\sqrt{3}$ and $1,\sqrt{3},\sqrt{4}$ and $1,\sqrt{4},\sqrt{5}$ follow. These dimensions give us our other squares, the Square Roots Squared, $\sqrt{1}\text{x}\sqrt{1}$, $\sqrt{2}\text{x}\sqrt{2}$, $\sqrt{3}\text{x}\sqrt{3}$, $\sqrt{4}\text{x}\sqrt{4}$, and $\sqrt{5}\text{x}\sqrt{5}$. These 10 squares and 5 triangles comprise our ***<u>Pythagorean Blocks</u>***, which you may purchase at ***http://mathfortheages*.com** or make them yourself.

1 x 1
Unit
Square

The measurements we have gleaned from the Unit Square return us to the Platonic Solids. We derive triangles formed by two radii and an external edge for each of the Polyhedra. We use these triangles to construct the solids from the inside out.

<u>The Extreme and Mean Ratio</u>: We finish with Kepler's precious jewel, the Golden Ratio, The Divine Proportion, Phi, ϕ.

Intro to the first edition:

Math for the Ages is a company dedicated to bringing mathematics to the world in a manner that does not stress drudgery. Many things are learned serendipitously through our inspiration and delight and not subject to direct instruction.

A knowledge of the basic fundamentals of Mathematics is critical to understanding the world in which we live. Math is at the center of modern technology. Math is at the center of Philosophy. Math is at the center of understanding statistics, economics, political science, physics, chemistry, biology; indeed it lies at the center of a lot of what it means to be a human being in the 21st century. It is a basic tool of understanding. It is a basic tool of invention.

This is not a new phenomenon. Mathematical understanding and skill lie at the heart of every great civilization in human history. All have contributed to the point at which we now find ourselves.

Do you know what a square root is? Can you explain the Pythagorean Theorem? What is an irrational number? If you cannot easily answer all of these questions then this book is for you. Or, even if you can, this book will be a useful companion in revisiting these fundamental building blocks.

This book is for people of all ages and abilities. It is an attempt to bring mathematics and its sister discipline geometry to everyone in a simple and deep manner. It takes the reader at their point of understanding and joins them in a leisurely stroll through some of the most basic and critical elements of mathematics. It is designed to be an entry point and guide to everyone from young children at play to college math students struggling to comprehend what may seem incomprehensible. It is also for the elderly who wish to understand and appreciate the beauty, integrity, and significance of a subject that may have been a mystery to them their whole life.

This is the first book from **Math for the Ages.**

Note: Both the original version and this updated version will, as with all of our other materials, remain a free download on the web. We encourage everyone to use these materials and share them with your friends. You can find them at http://mathfortheages.com

In a Circle

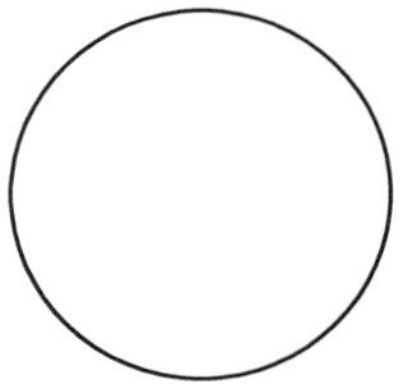

Mathematics begins with the circle. Our world is dominated by two heavenly circles, the sun and the moon. The sun and the moon are not only the most significant shapes in the sky, they are the most significant shapes in our cultural and mythological lives as well.

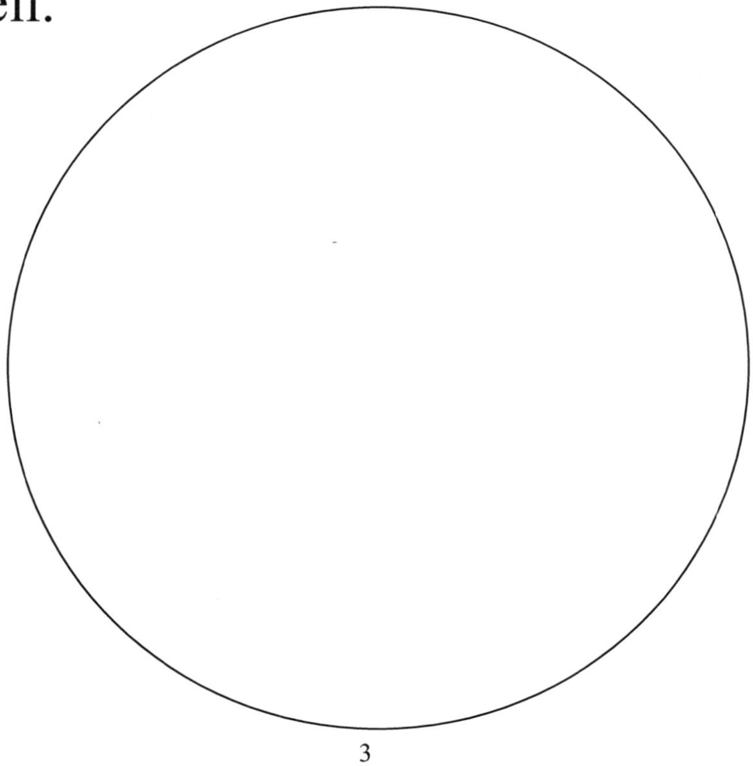

Circular Logic, Making Pi

At Math for the Ages almost everything begins with the circle. It is the most elegant and efficient of nature's forms.

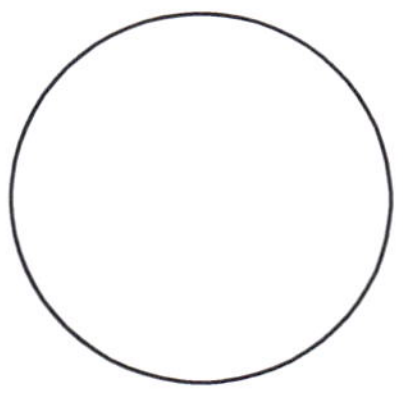

The circle has two major parts, the Focus and the Locus. The Focus is the center point. The Locus is all of the points that form the circle itself. The definition of the circle is, the locus of points equal distance from a single focus.

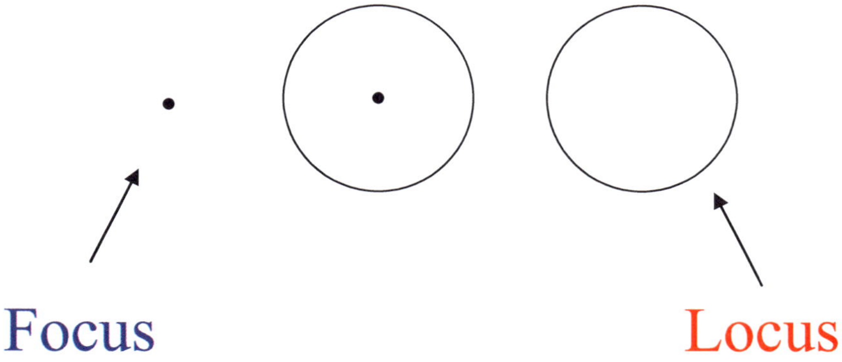

The Circle has four main measures.
The Radius, the Diameter, the Circumference, and the Area. The Radius is the distance from the Focus to the Locus. The Diameter is the distance from a point on the circle to another point on the circle that is directly opposite, and passes through the Focus.

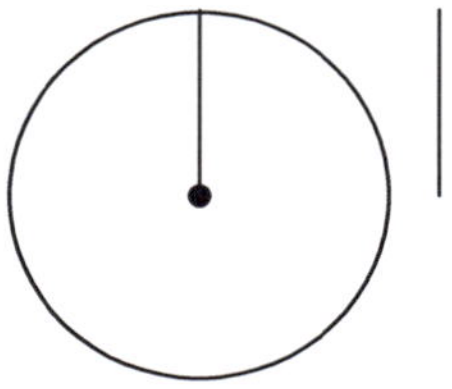

Radius = $\frac{1}{2}$

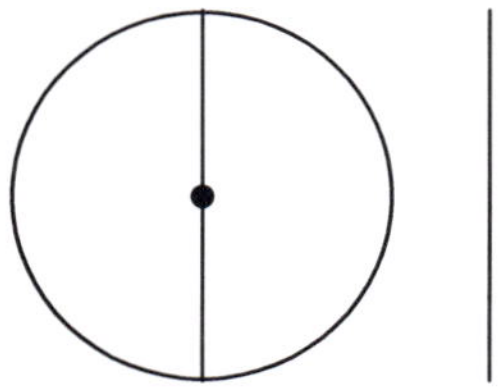

Diameter = 1

Two Radii are equal to one Diameter $2R = D$
The circumference is the distance around the circle.

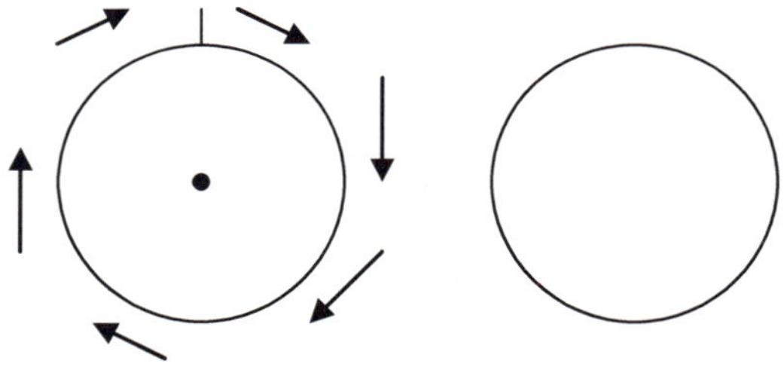

Circumference = πD
π, Pi, is an Irrational Number
$\pi \approx 3.14159265...$

The Ratio of the Diameter to the Circumference is Pi. If the Diameter is one, the circumference is Pi or π. $C = \pi D$ If $D = 1$ then $C = \pi$

Area $= \pi R^2 = \pi(\frac{1}{2})^2 = \pi/4 \text{ in}^2$

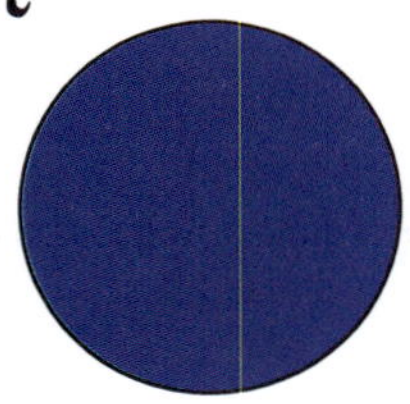

Hocus Pocus Focus Locus - Making a Circle

Circle Maker
Focus Stick

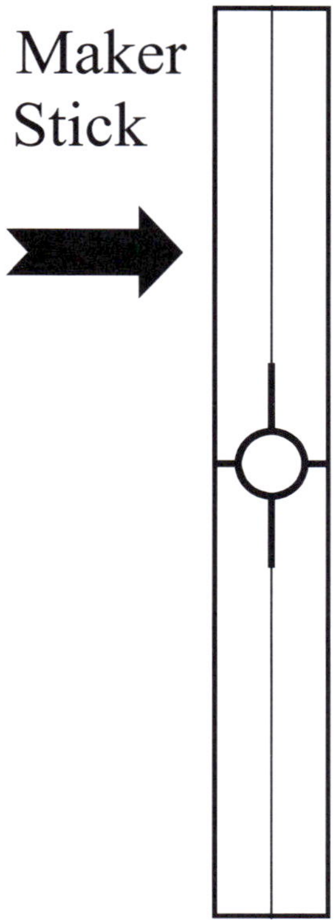

Mathematics begins with the circle and the circle begins with the focus/center point. We use this focus/center point and our tool, the circle maker, to build the circle.

Focus

The circle maker/focus stick may be made of wood or a thick piece of paper like card stock. It is a thin rectangle with a hole in the exact center. The focus/center point is a dot on a piece of paper.

Our circle maker shown here is printed on card stock.
4 inches by 3/8 inch.

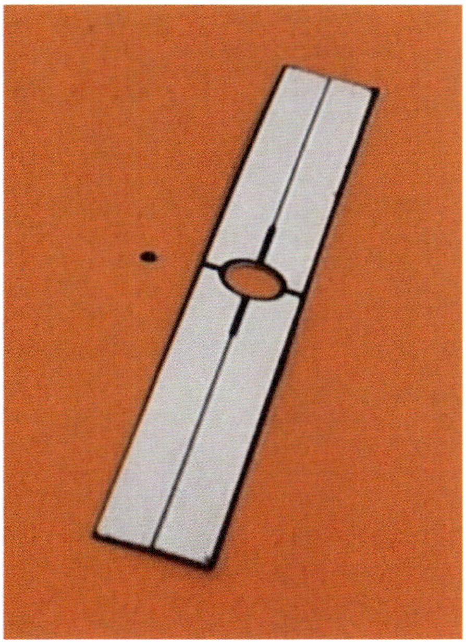

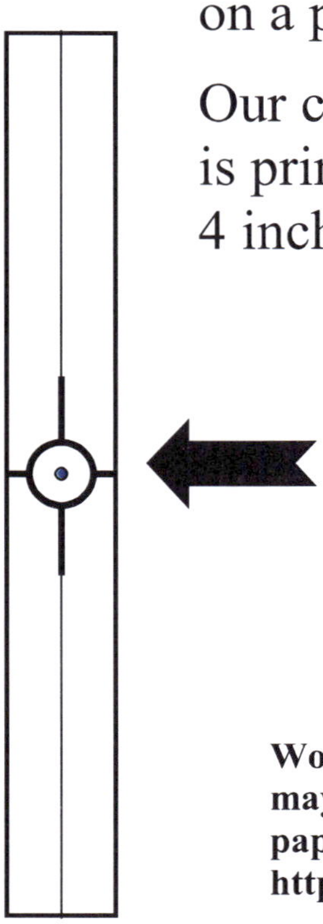

Place the circle maker hole directly over the focus

Wooden Circle Makers/Focus Sticks may be purchased and the pattern for paper versions may be downloaded at http://mathfortheages.com

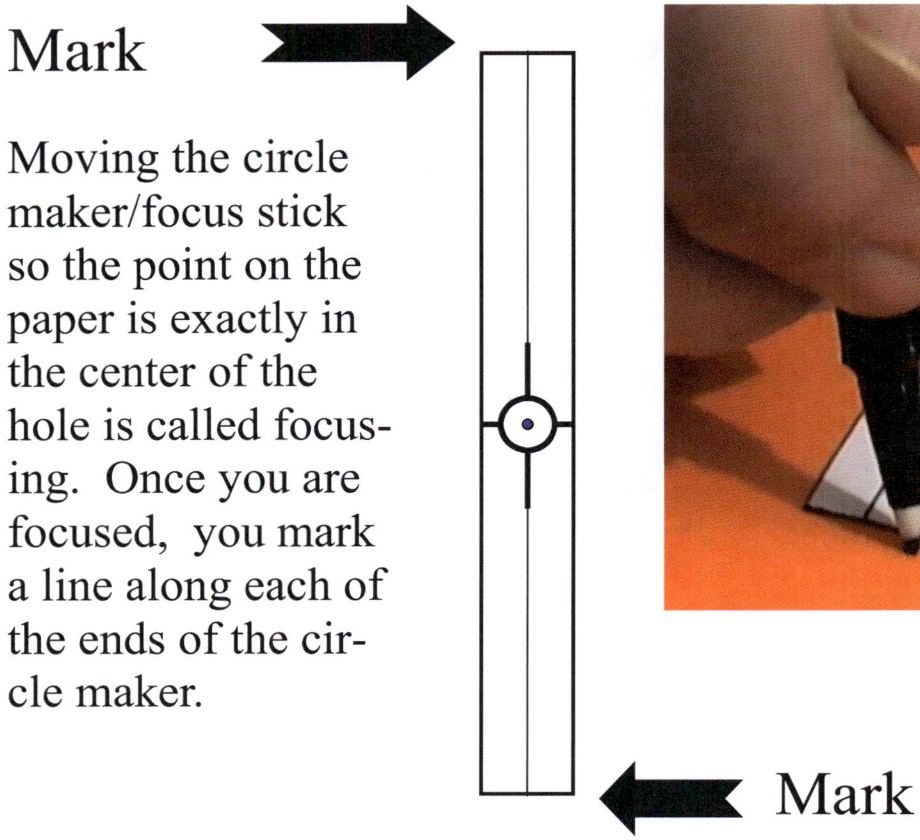

Moving the circle maker/focus stick so the point on the paper is exactly in the center of the hole is called focusing. Once you are focused, you mark a line along each of the ends of the circle maker.

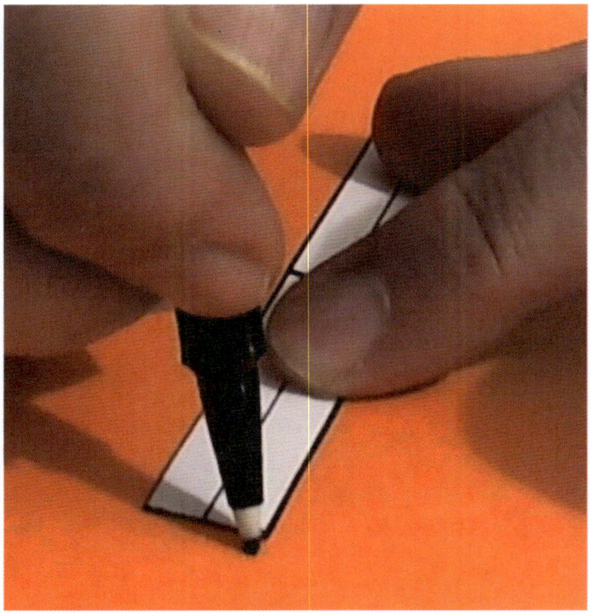

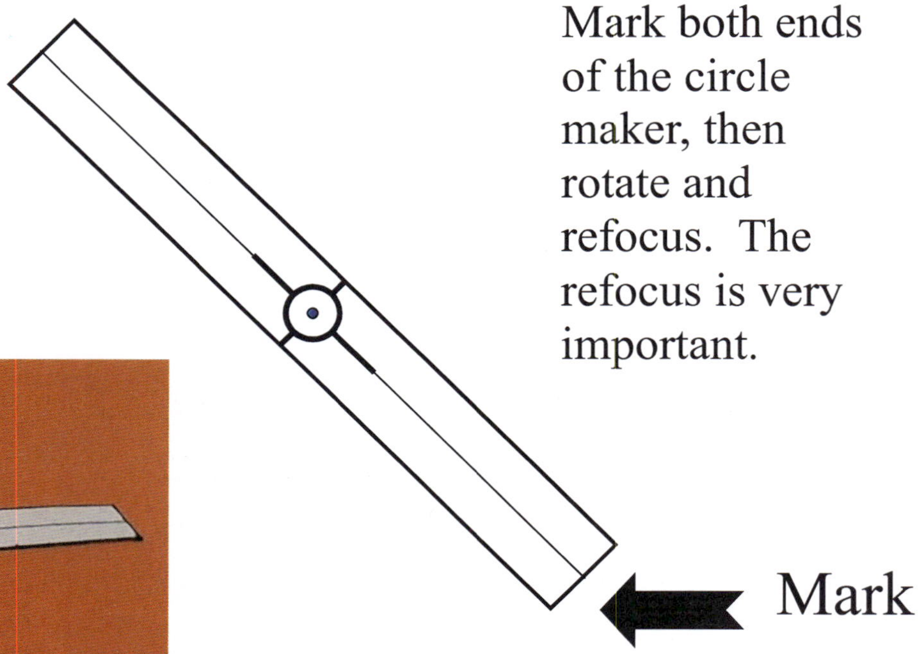

Mark both ends of the circle maker, then rotate and refocus. The refocus is very important.

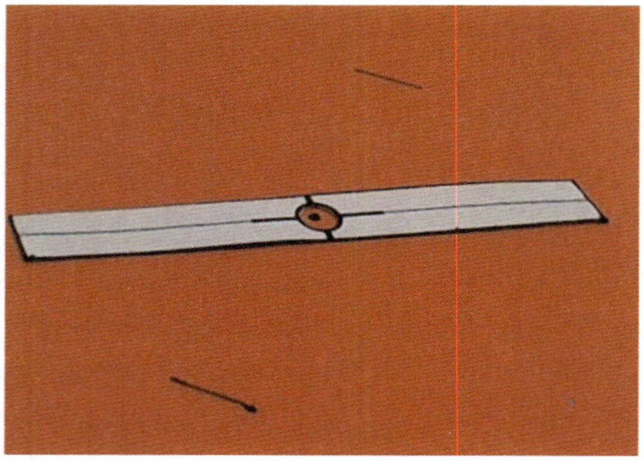

Continue: Move the Circle Maker, refocus and mark.

The point on the paper must be kept directly in the center of the circle maker's hole. Once you are refocused, mark the ends again.

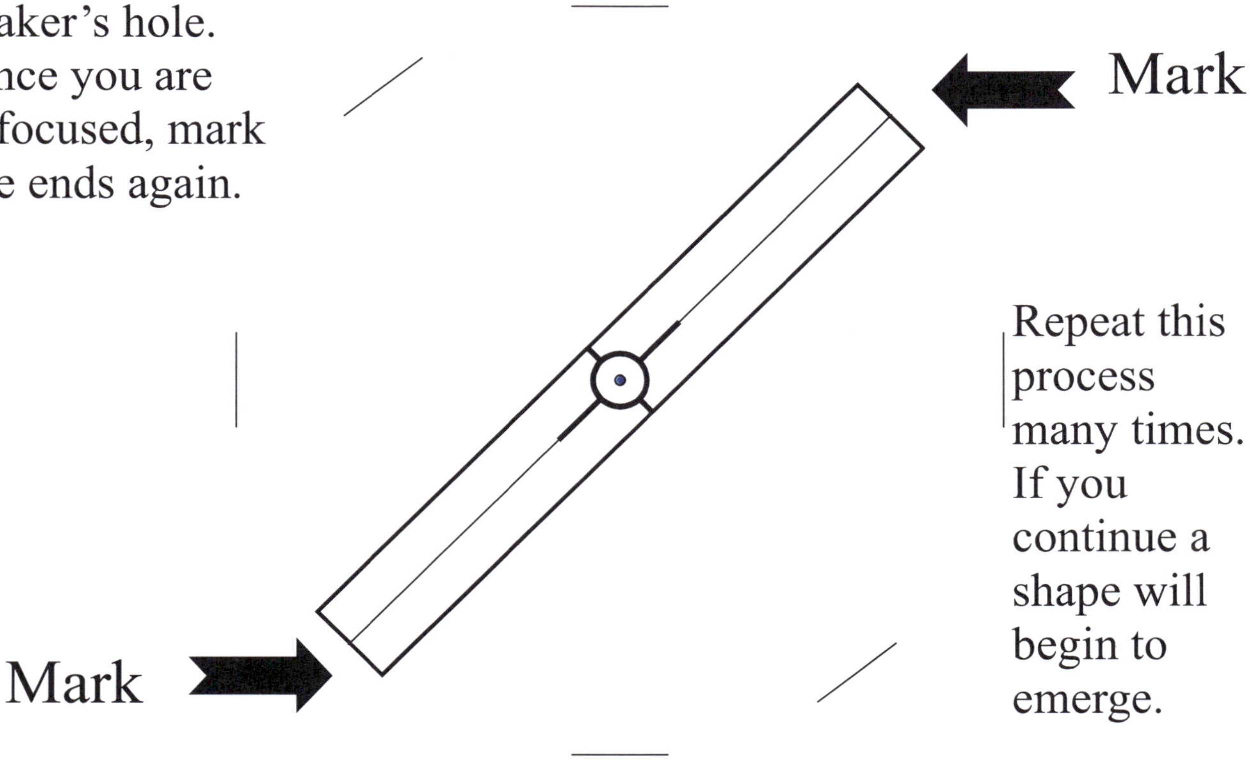

Repeat this process many times. If you continue a shape will begin to emerge.

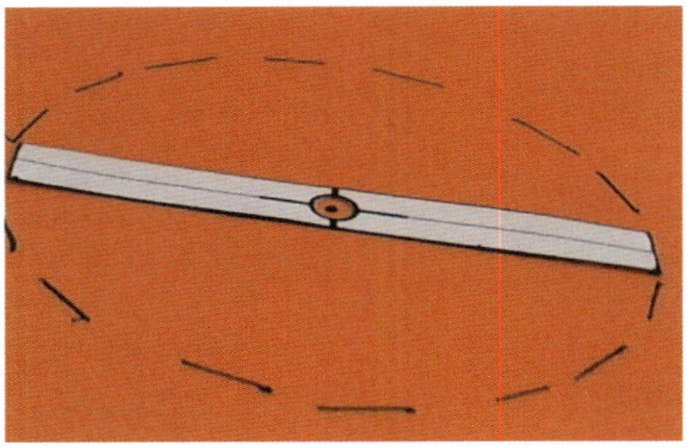

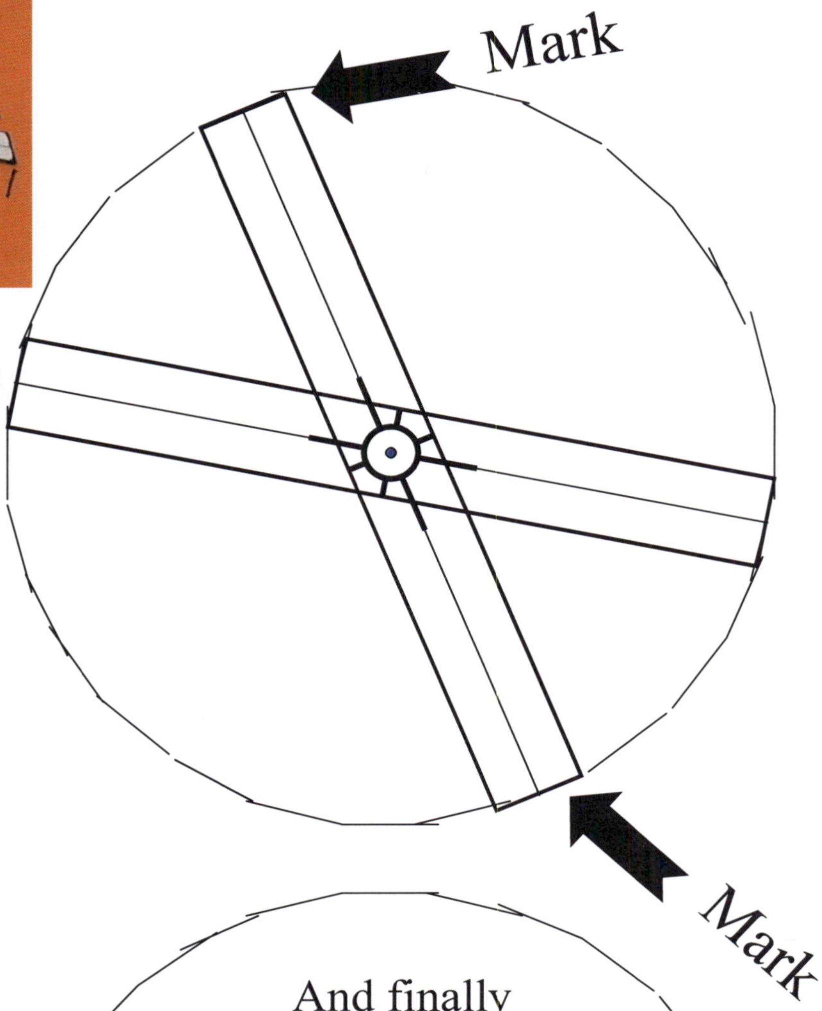

Fill in all of the gaps by rotating the circle maker to the vacant spots and marking a line at the end of the circle maker.

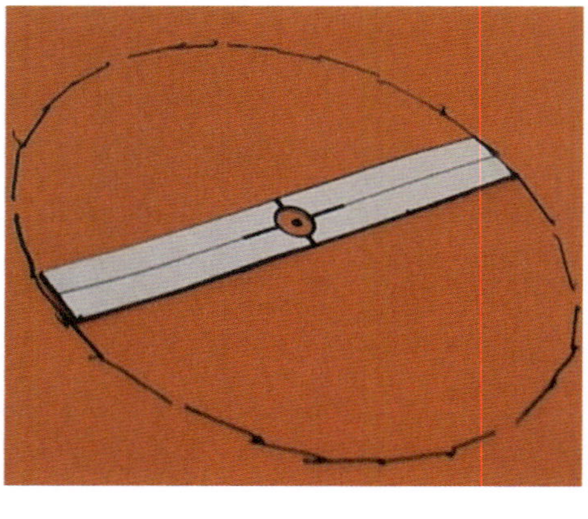

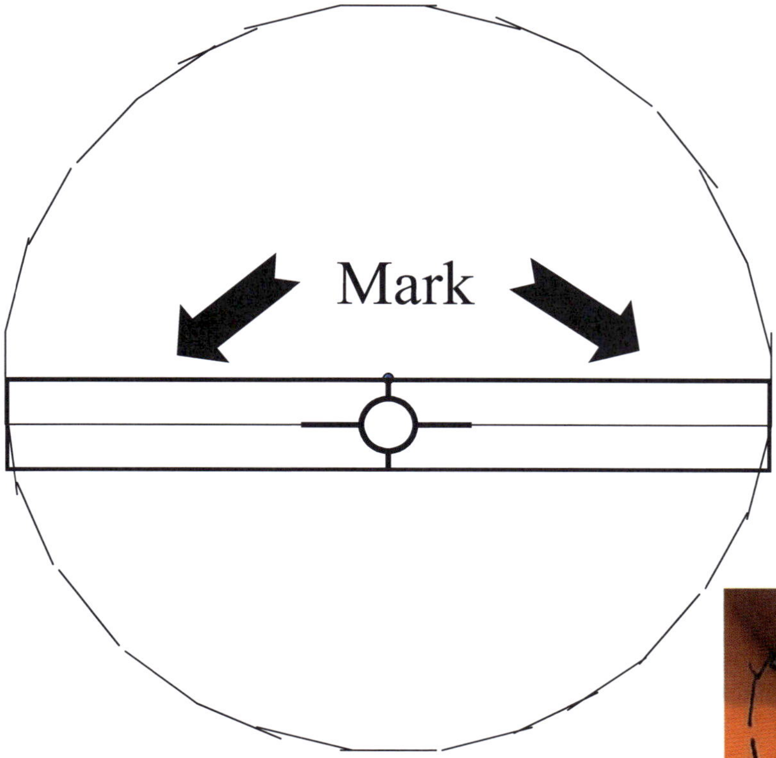

Using the circle maker as a straight edge, draw a line that goes from one point on the circle to a point on the other side of the circle This line is called a chord. Pass this chord through the focus.

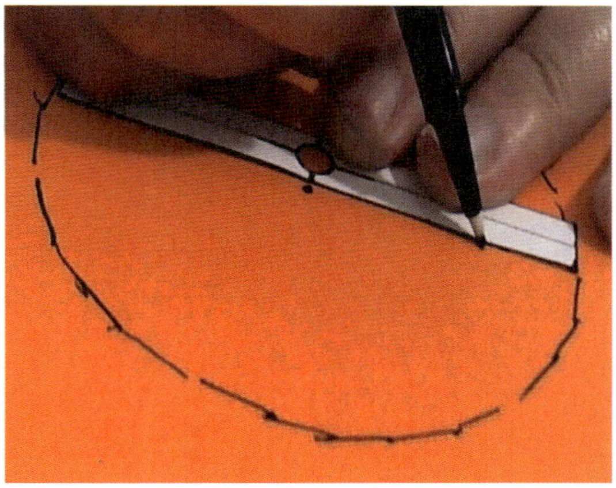

This special chord is called the

Diameter

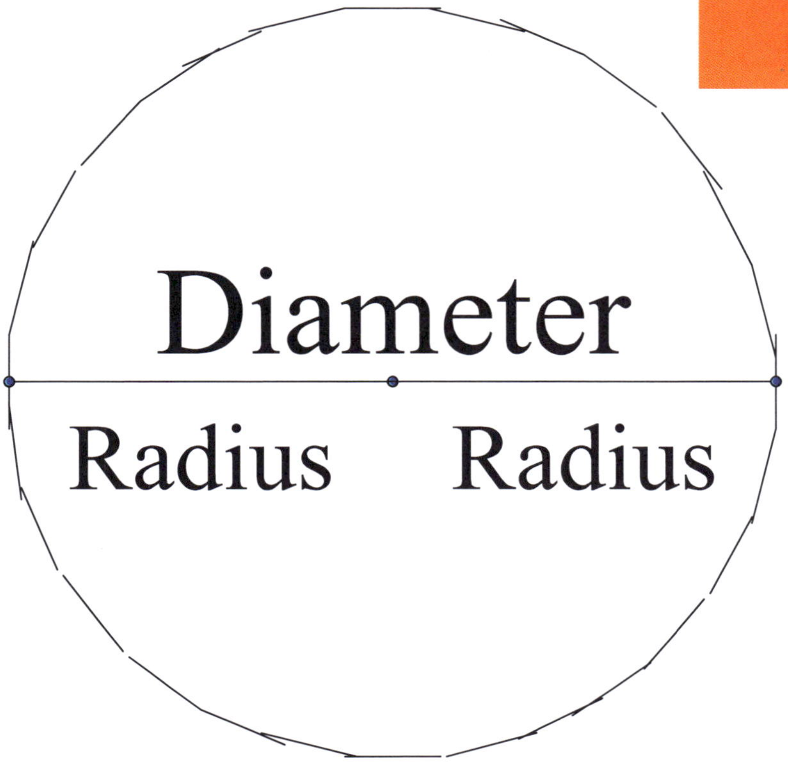

One half of the Diameter, the line that goes from the focus to a point on the circle is called the

Radius

Mark a point where the diameter line meets the circle.

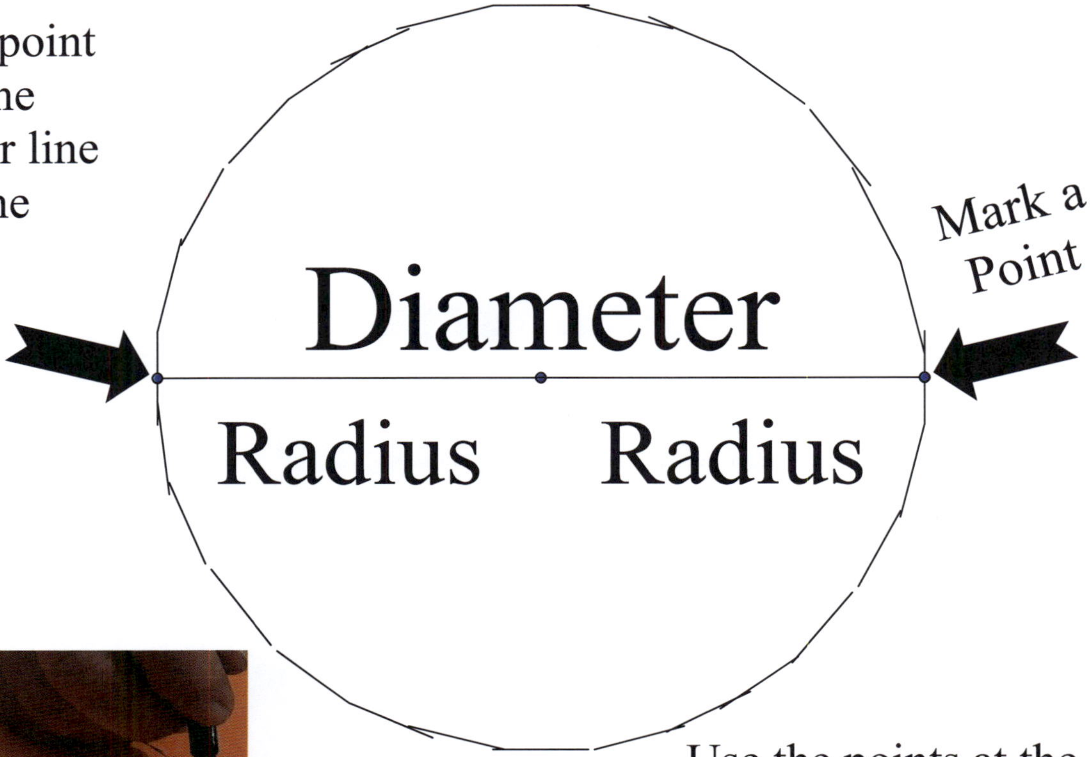

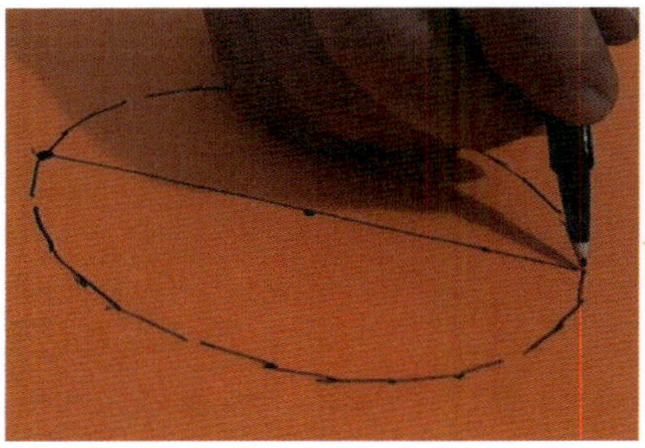

Use the points at the end of the diameter lines as the focus for the circle maker.

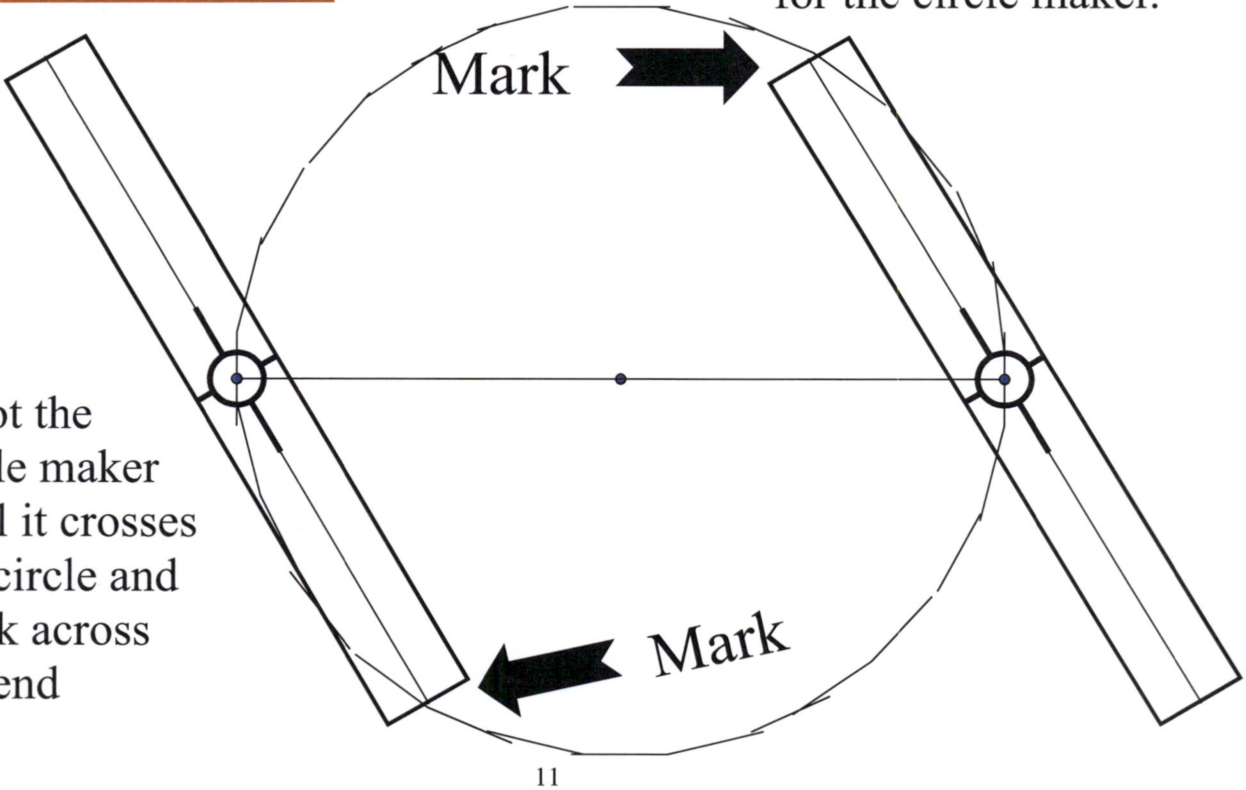

Pivot the circle maker until it crosses the circle and mark across the end

Pivot the circle maker to the other sides and mark them as well.

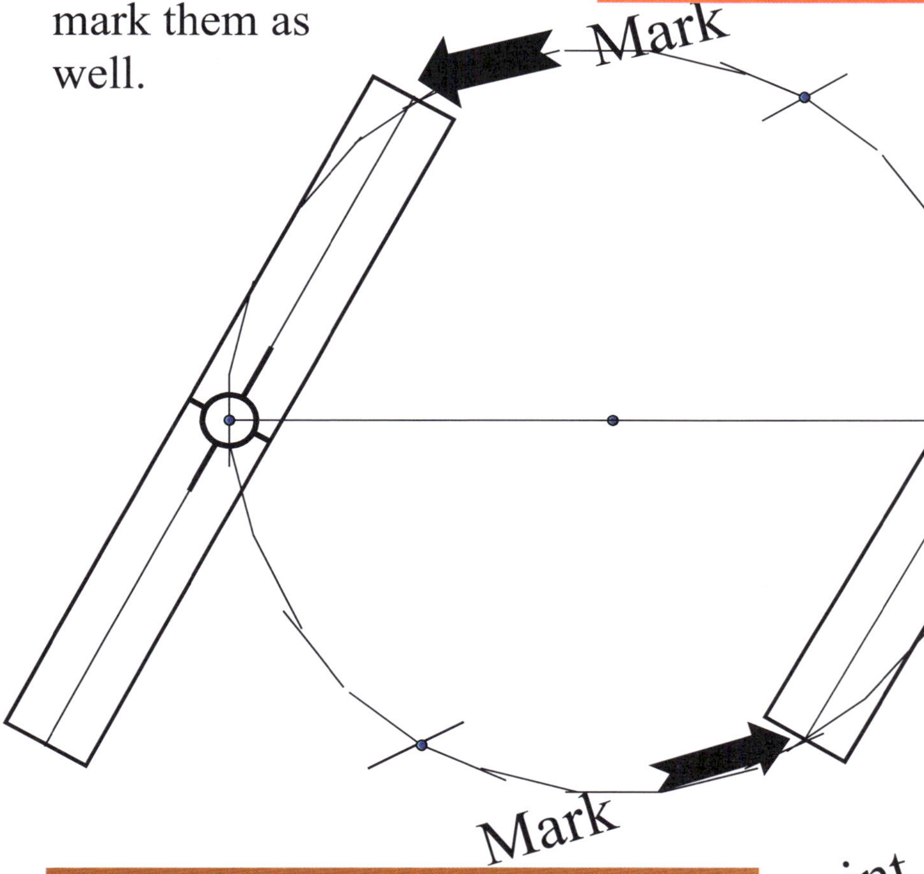

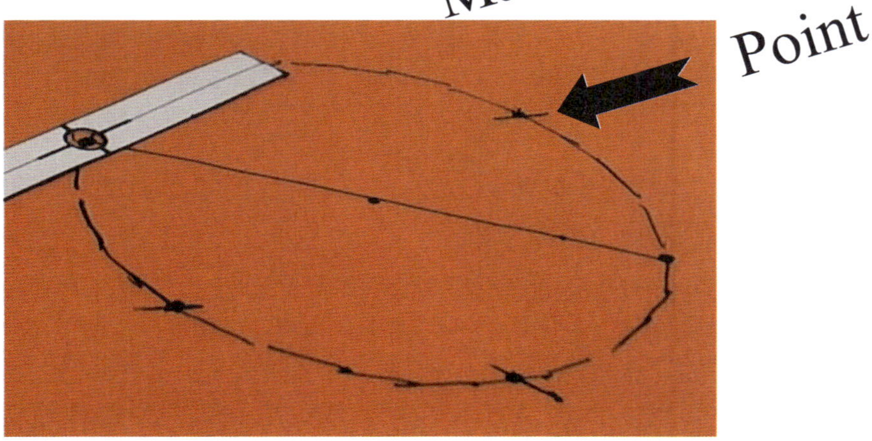

At each location that these lines cross the circle, mark a good solid point.

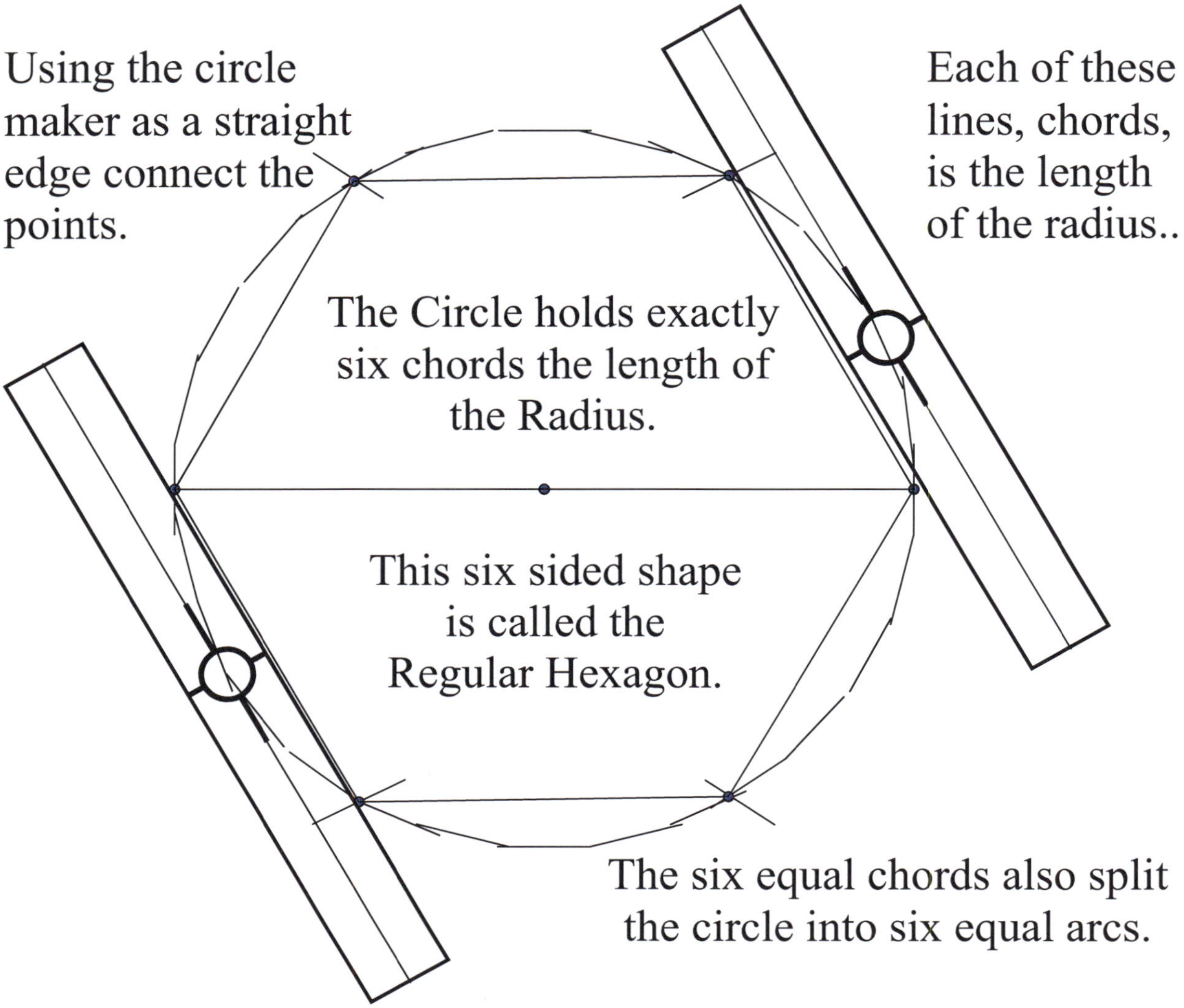
Using the circle maker as a straight edge connect the points.
Each of these lines, chords, is the length of the radius..
The Circle holds exactly six chords the length of the Radius.
This six sided shape is called the Regular Hexagon.
The six equal chords also split the circle into six equal arcs.

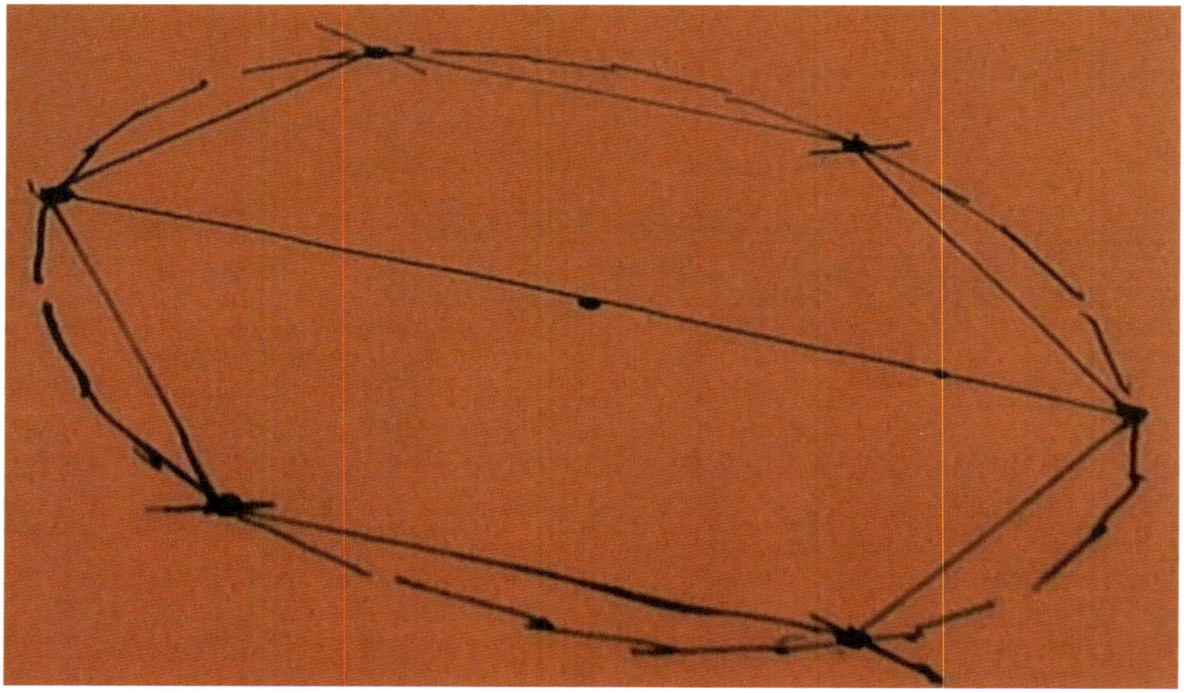

Using the circle maker as a straight edge, connect each of the points to the points opposite them on the circle.

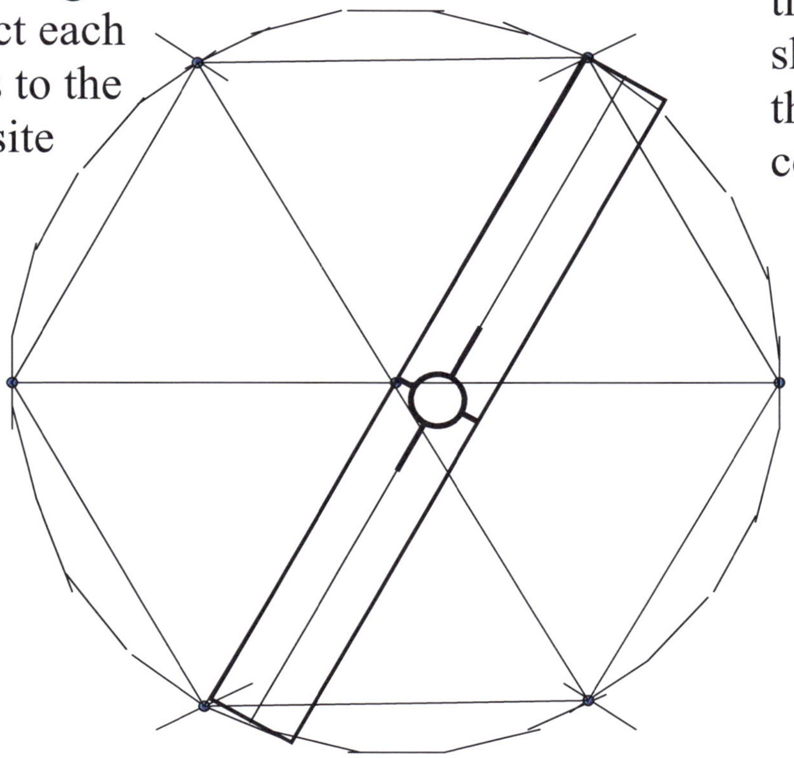

Each of these chords should pass through the center.

These two chords are Diameter Lines..

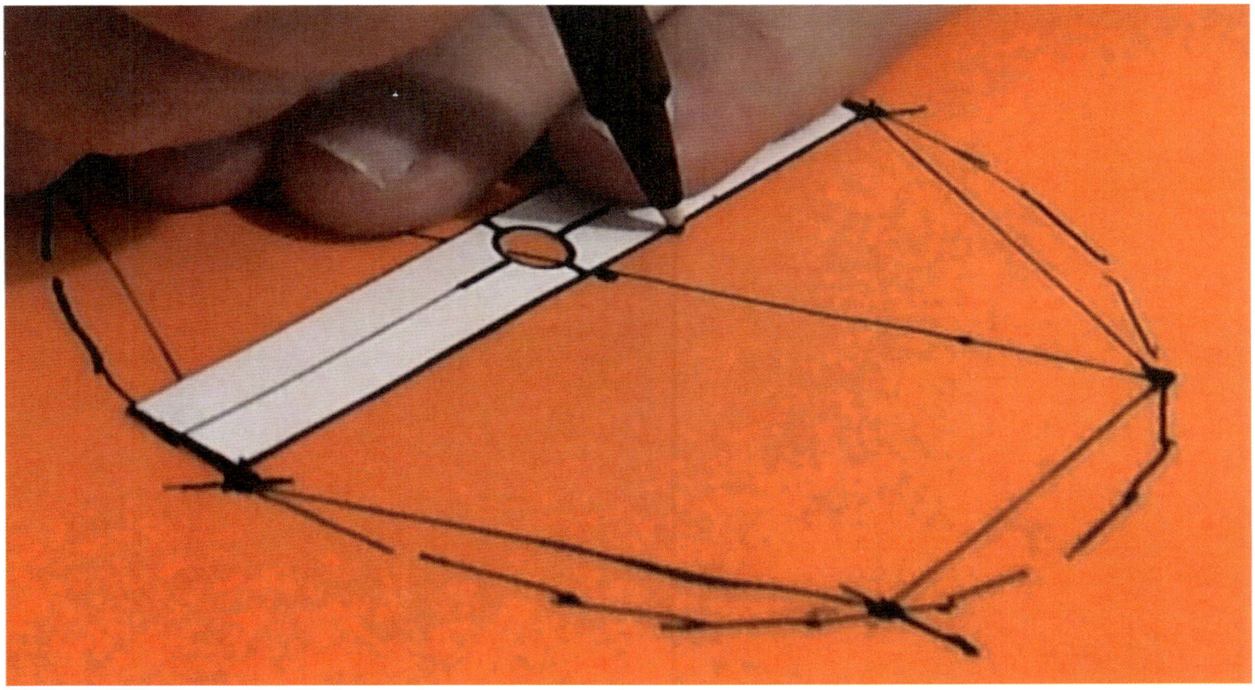

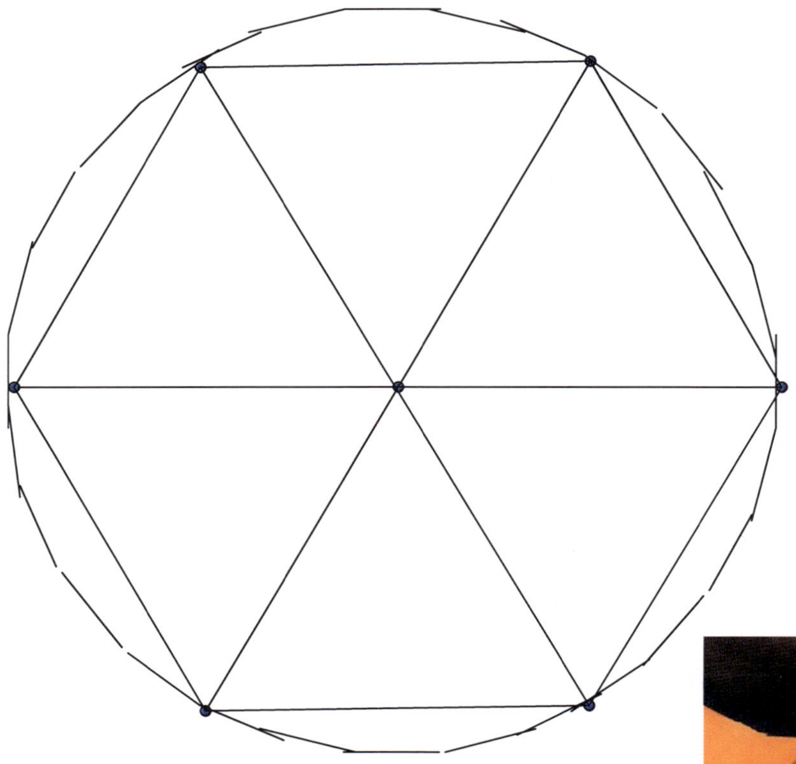

This leaves us with a Hexagon inscribed in a circle. This hexagon is a regular hexagon. All of the sides and the angles are the same. The regular hexagon is made up of six regular triangles.

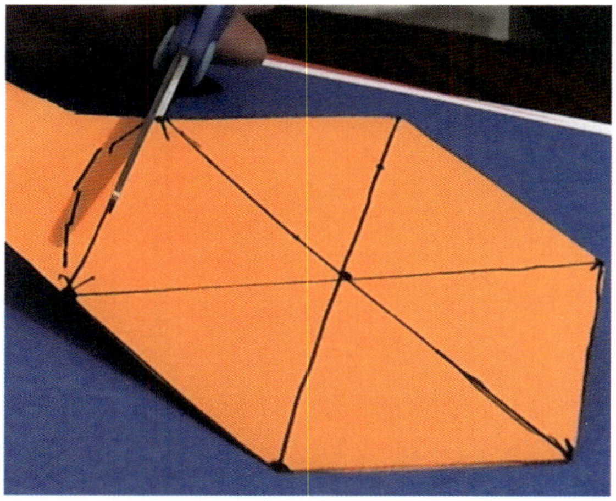

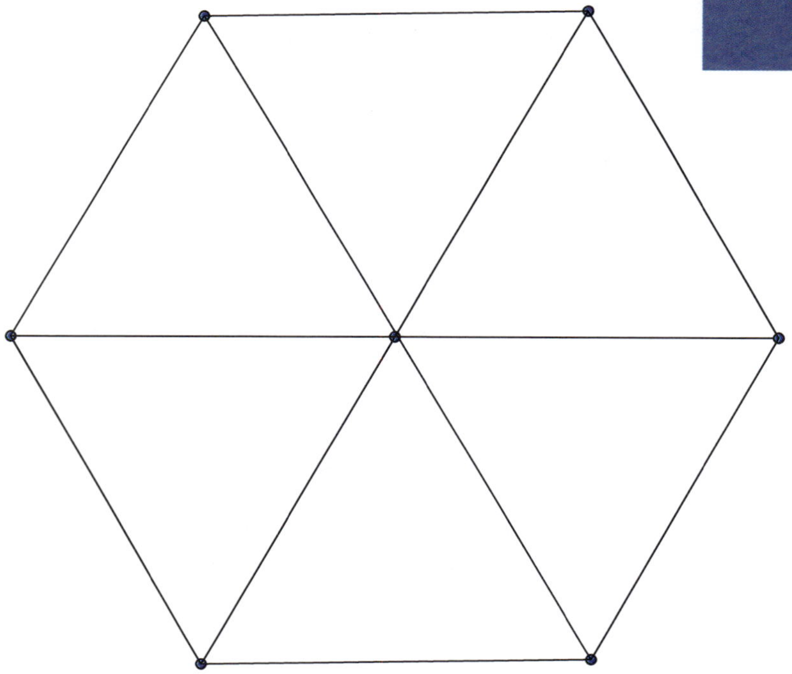

We trim away the circle by cutting along the perimeter of the hexagon. The edges and the radial lines are the same length. The regular hexagon divided into six regular triangles is the key to constructing many other shapes.

Fold and crease each of the diameter lines of the hexagon. This leaves each of the triangles with folded edges on the their sides. Cut one of the radial lines from the edge to the center of the Hexagon.

Fold under one of the triangles and clip it in place with a paper clip. This leaves us five triangles around a point that when pressed down on a flat surface forms a three dimensional structure whose outer edge is a Pentagon.

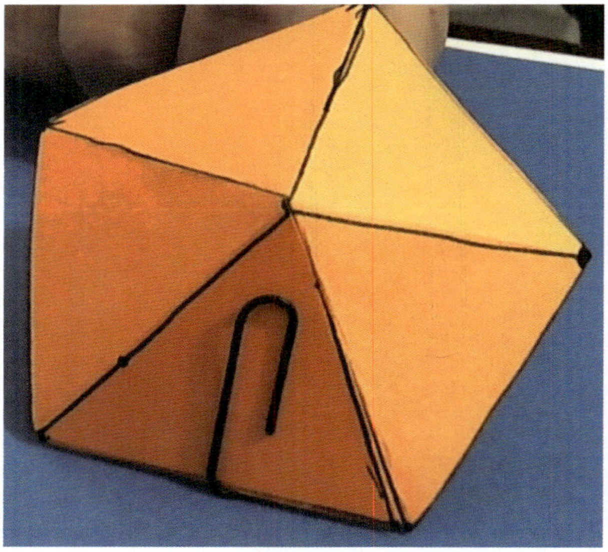

Trace the Pentagon.

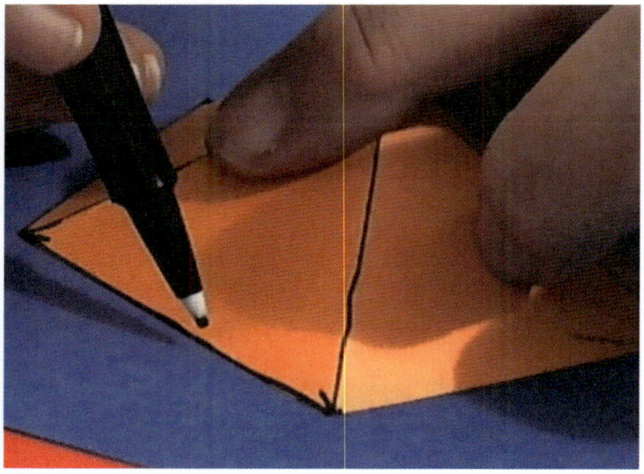

Note the configuration: five triangles around a point, is exactly the exterior arrangement of the faces of the Icosahedron. The Pentagon is the face of the Dodecahedron.

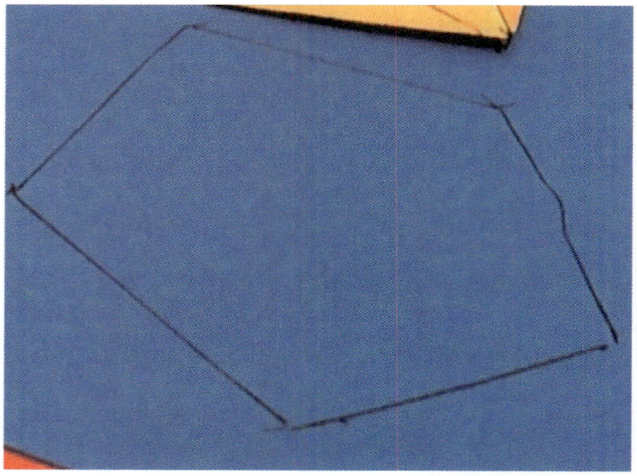

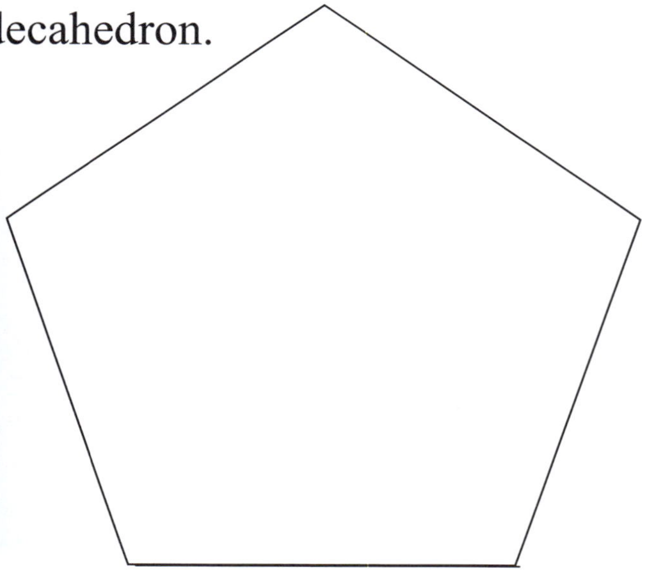

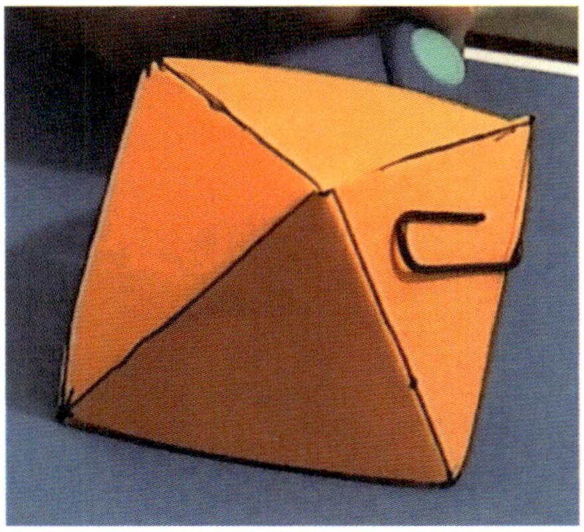

Fold two of the triangles under and clip with a paper clip. This leaves us four triangles around a point that when pressed down on a flat surface forms a three dimensional structure whose outer edge is a Square. Press down and trace. This triangle configuration is identical to that of the exterior of the Octahedron. The Square is seen again as the face of the Hexahedron, the Cube.

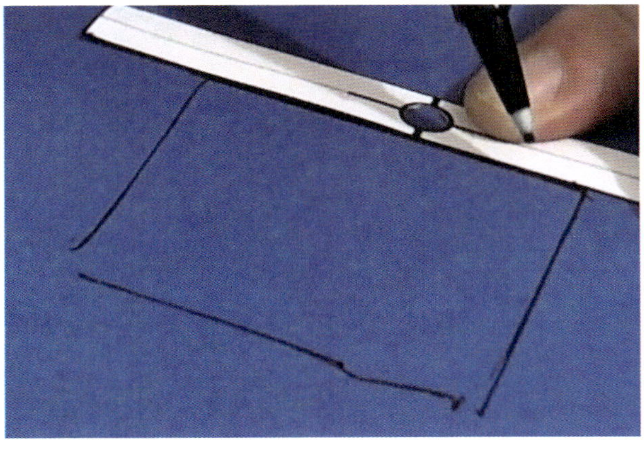

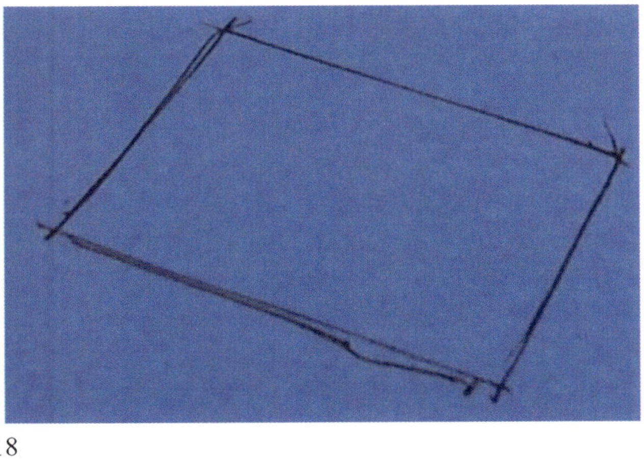

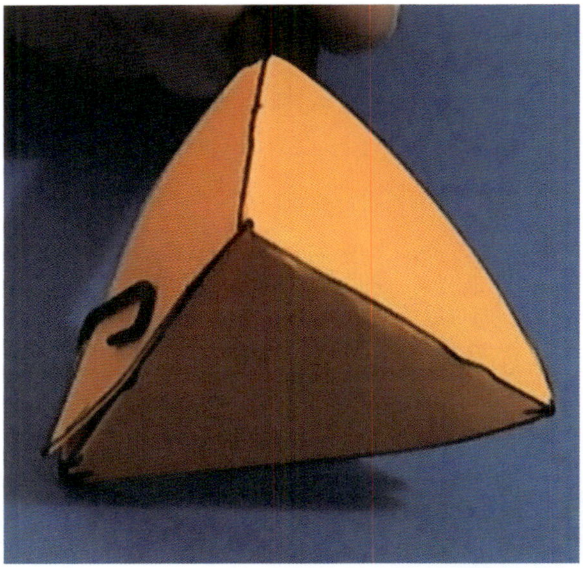

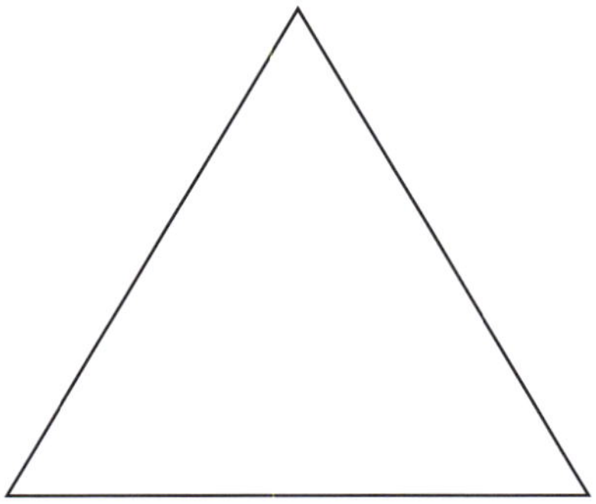

Tuck three of the triangles under and clip. This gives us three triangles around a point that when pressed down gives us another triangle. Press down and trace. Note: This configuration of the triangles is the exact configuration of the exterior of the triangle faces of the tetrahedron and the triangle created is equal to the other three.

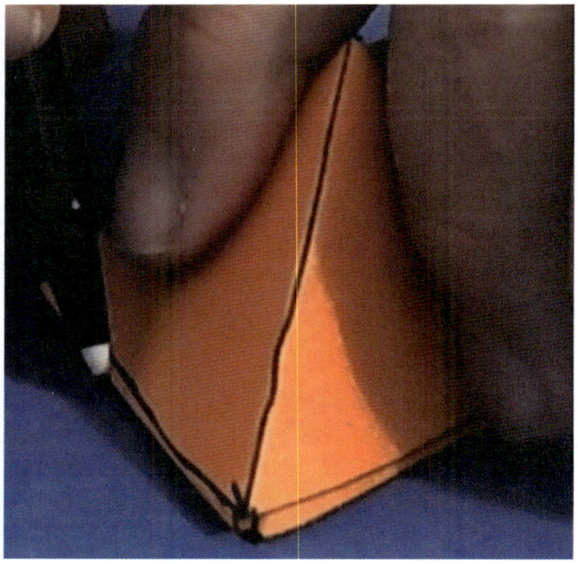

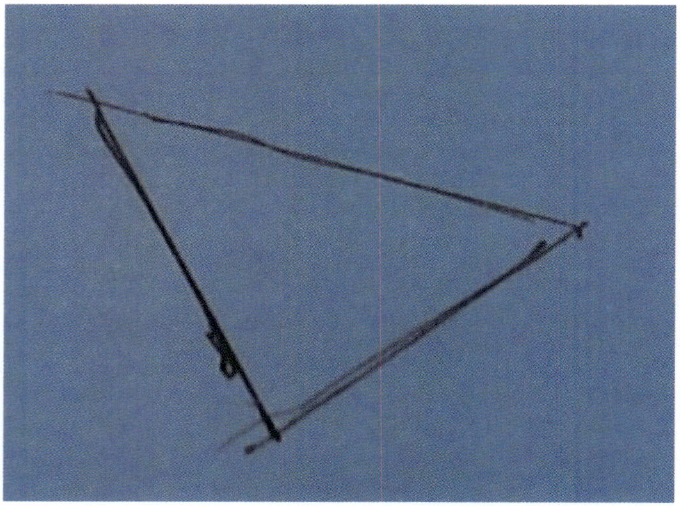

Thus, beginning with the circle we get the hexagon, and from the hexagon we derive the pentagon, square and triangle.

Circular Logic
A Matter of Degrees

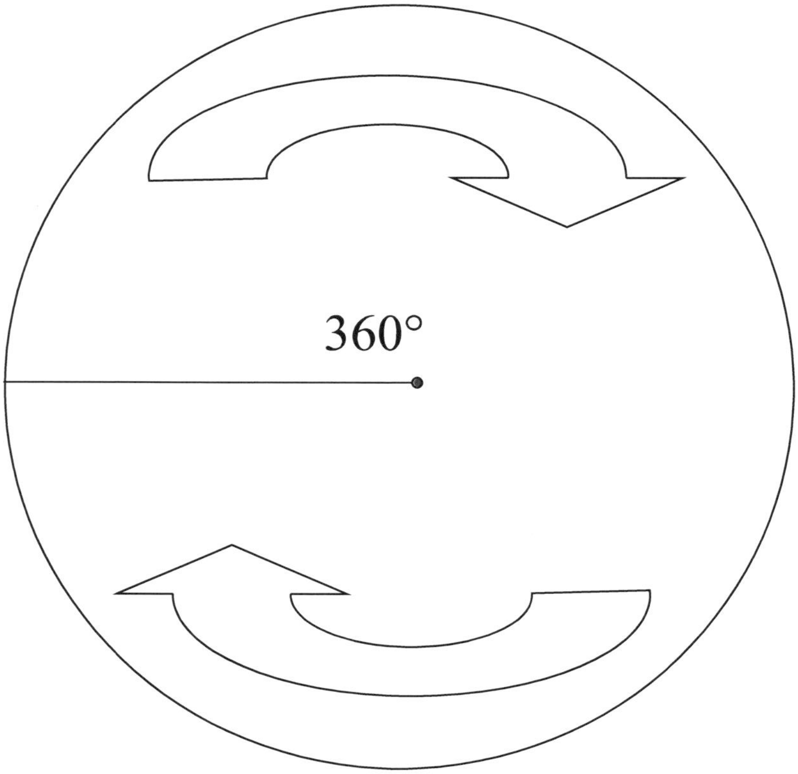

The Circle gives us the notion of Degrees. There are 360 Degrees in a Circle. We use degrees to measure angles.

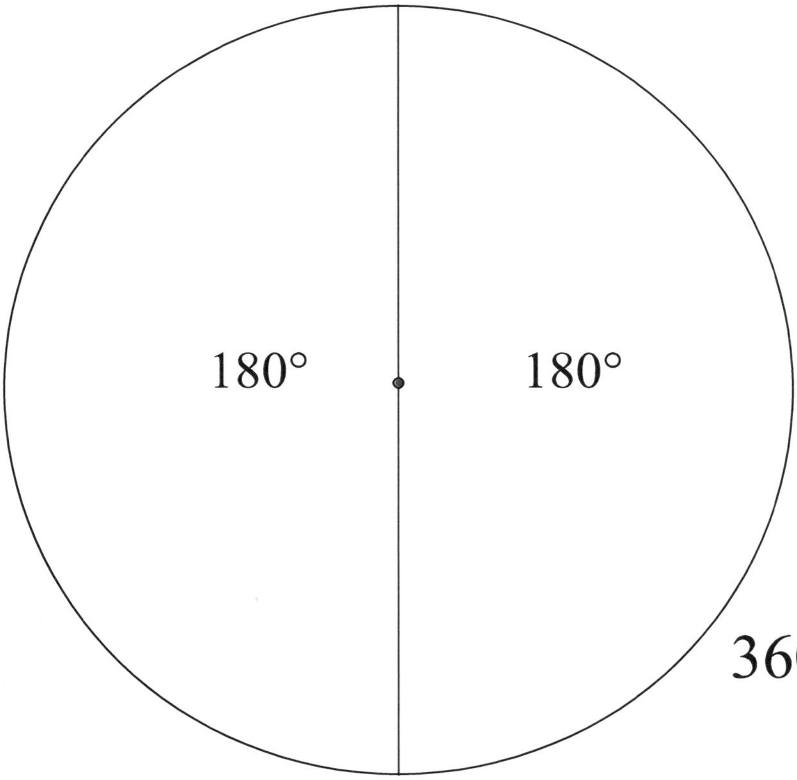

We can divide the circle in two parts. Half a Circle is 180 Degrees. This is a straight line.

$$360/2 = 180$$

We take the fact that a circle has 360°and add to that the fact that in a triangle, any triangle, the three angles add up to 180°. The center angle of each triangle is 60°. This leaves 120° for the other two. The other two angles are equal which means that they are each 60°. Three equal angles of 60°.

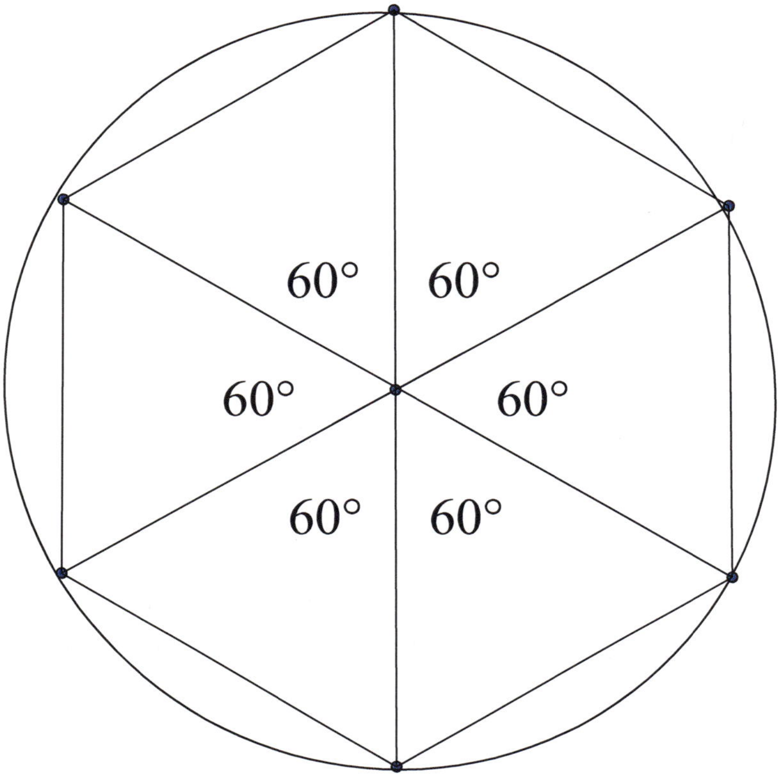

Divide the center into 6 equal parts.
360/6 = 60

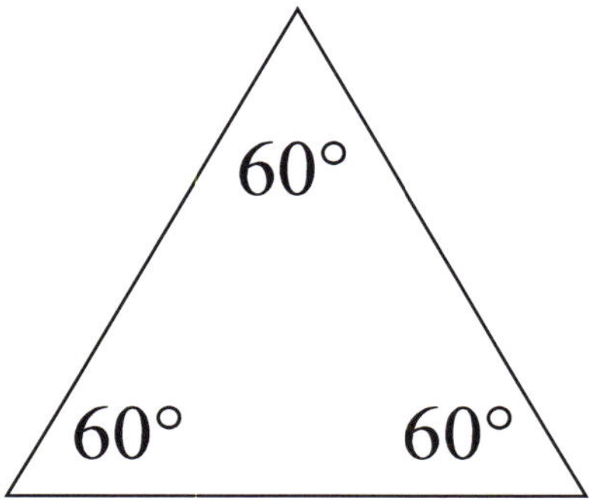

The Regular Triangle

When these triangles are put together two of the 60° angles come together to form 120° angles for the hexagon.

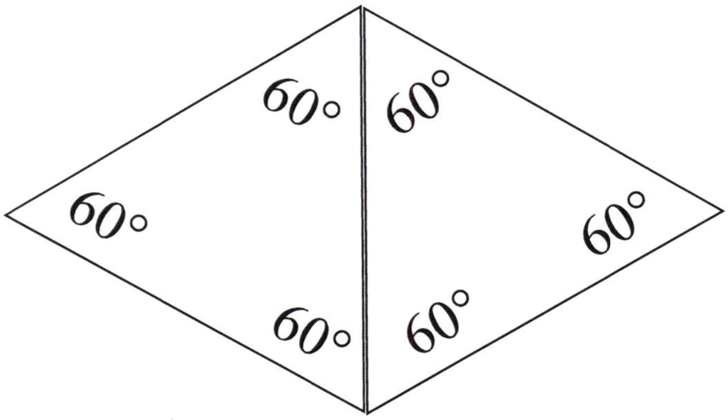

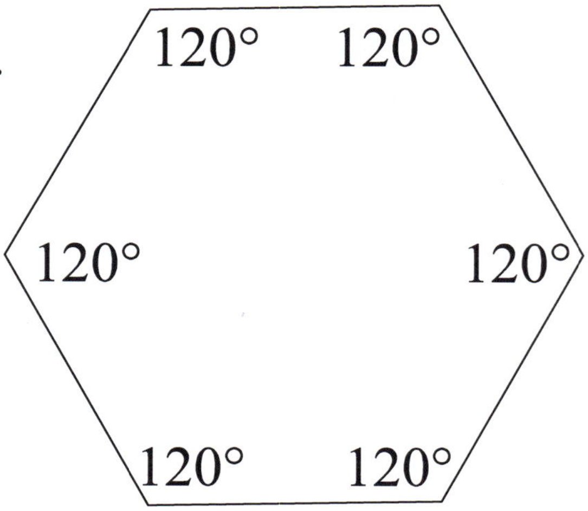

We now come to the pentagon we have constructed. We can make 5 triangles that come together in the center. We divide the 360° in the center by 5 which gives us 5 angles of 72° each. This leaves 108° for the other two angles which we know are equal. They have 54° each.

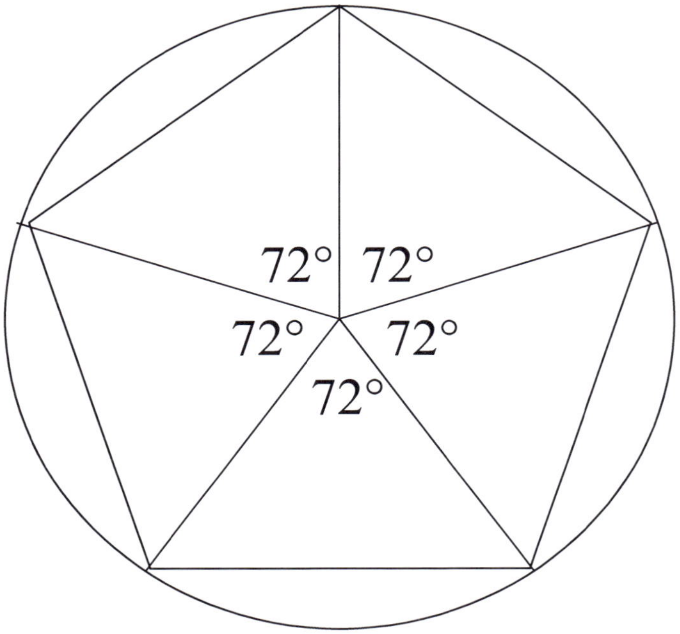

$$360/5 = 72$$

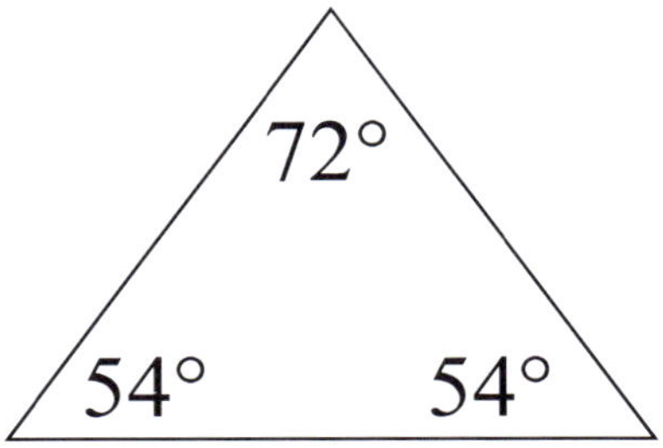

The Pentagon Component Triangle

Five of these triangles together make the pentagon. Two of the 54° angles touch each other in each of the five corners of the pentagon. The corners of the pentagon are thus 108°.

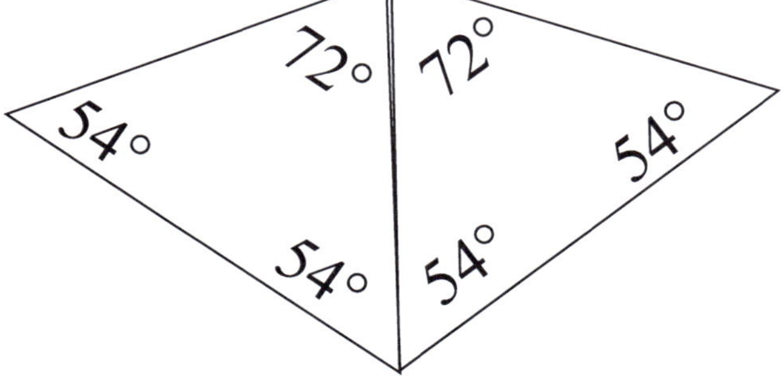

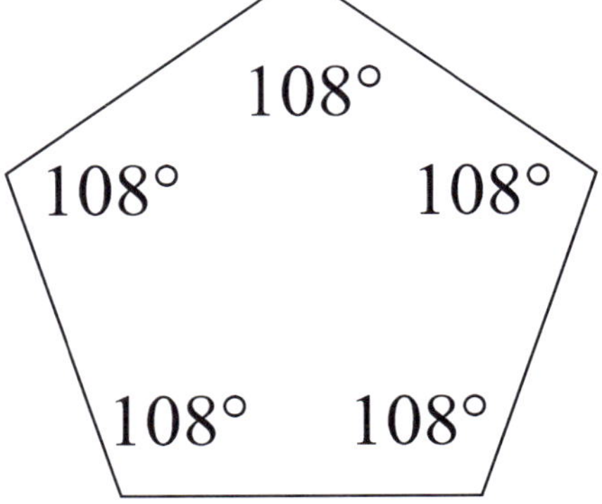

360/4 = 90

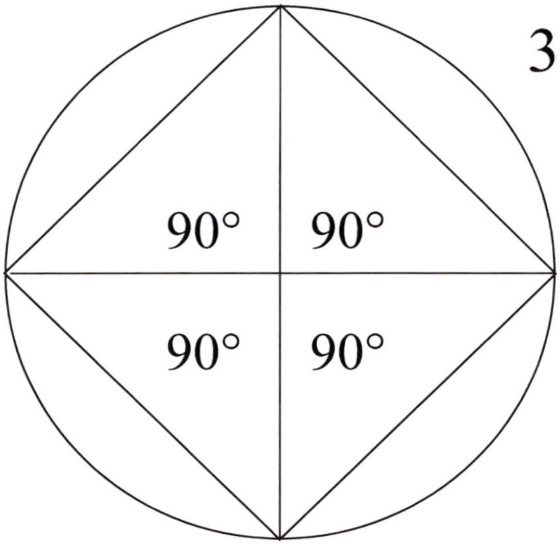

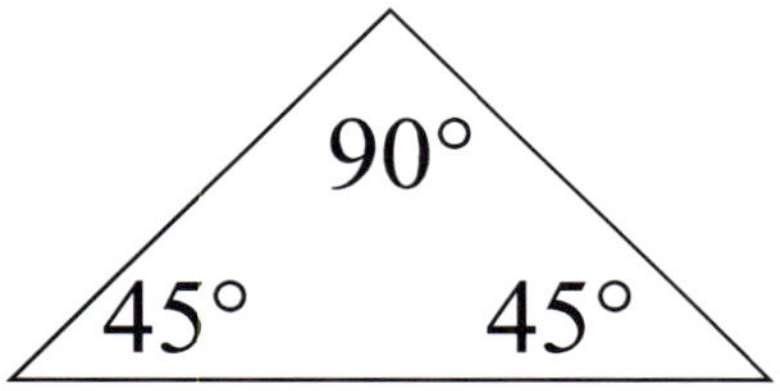

Divide the square into four equal angles of 90°. This leaves 90° for the other two angles, 45° each.

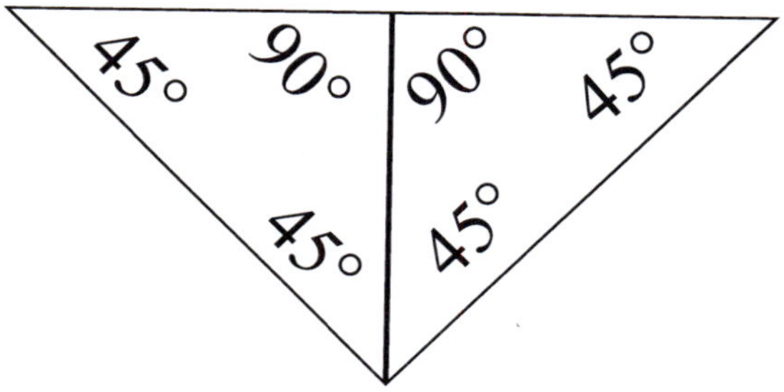

The two 45° angles meet at each corner making the corners of the square 90°

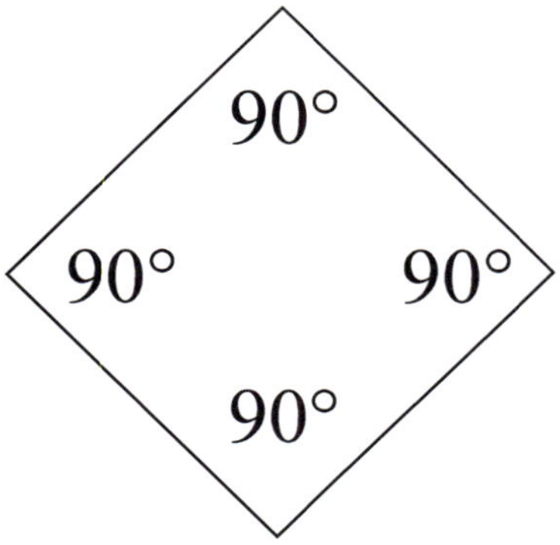

360/3 = 120

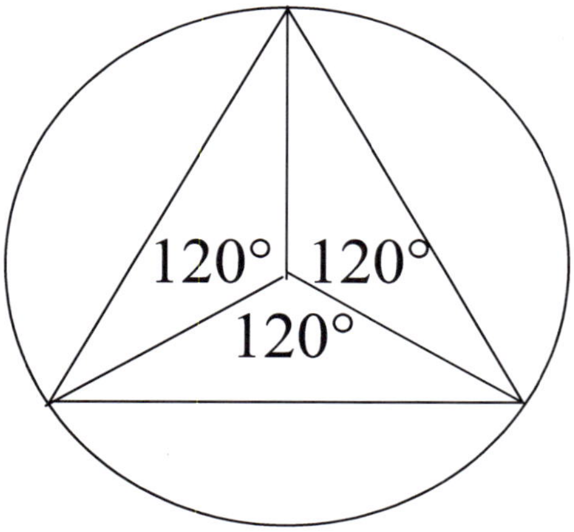

The triangle splits into three 120° angles leaving 60° for the other two, or 30° each. These angles meet at each corner which recreates our regular triangle of 60° corners.

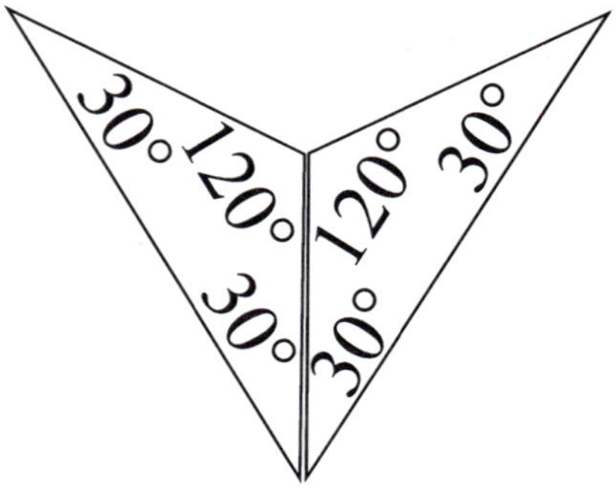

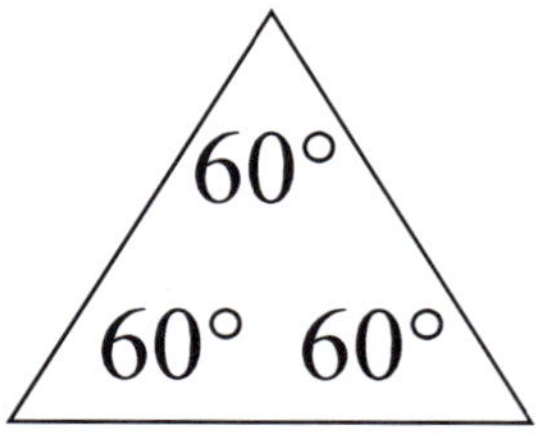

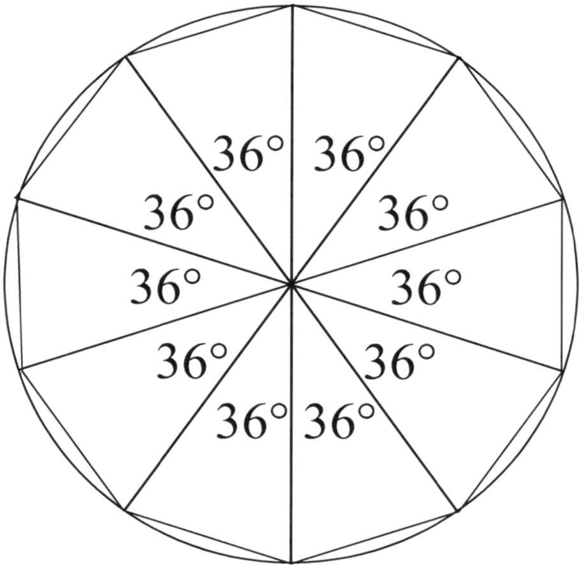

360/10 = 36

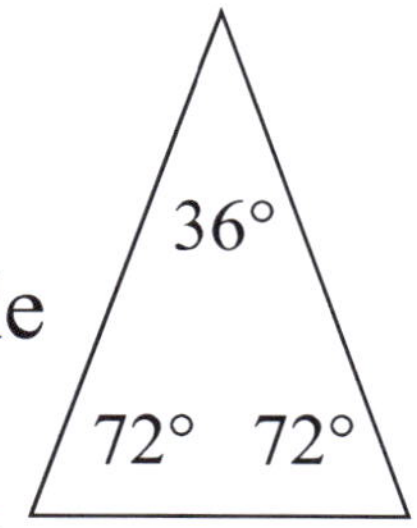

We will see this triangle again. It is called the <u>Golden Triangle</u>. If its base is 1 its sides are ϕ. Ten of them make a Decagon.

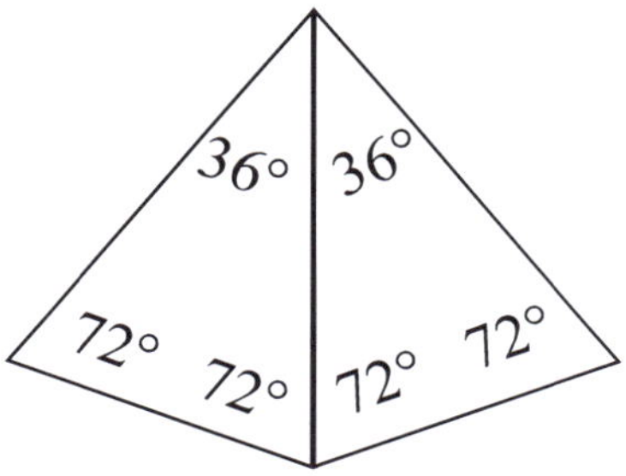

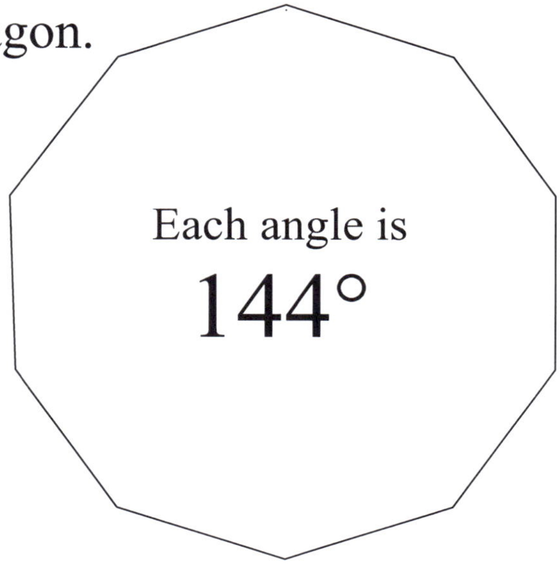

The earth is divided into 360° and 24 time zones. Each time zone has 15°. Geometry means measure the earth.

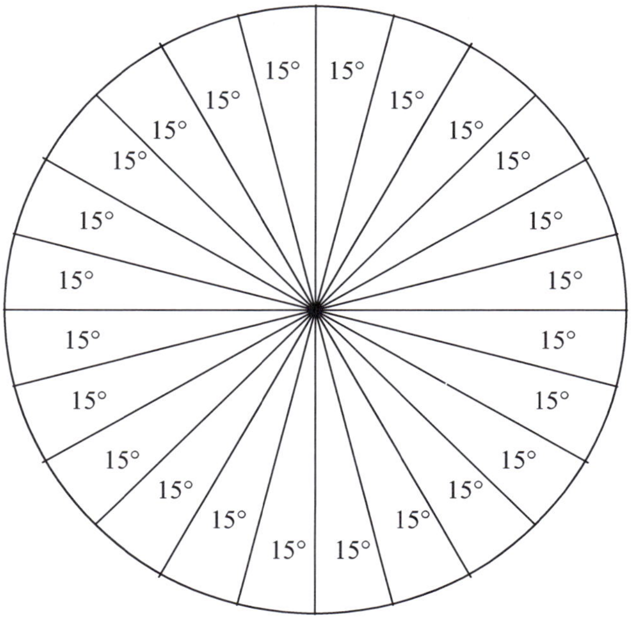

The Exterior Faces
Of the
Platonic Solids

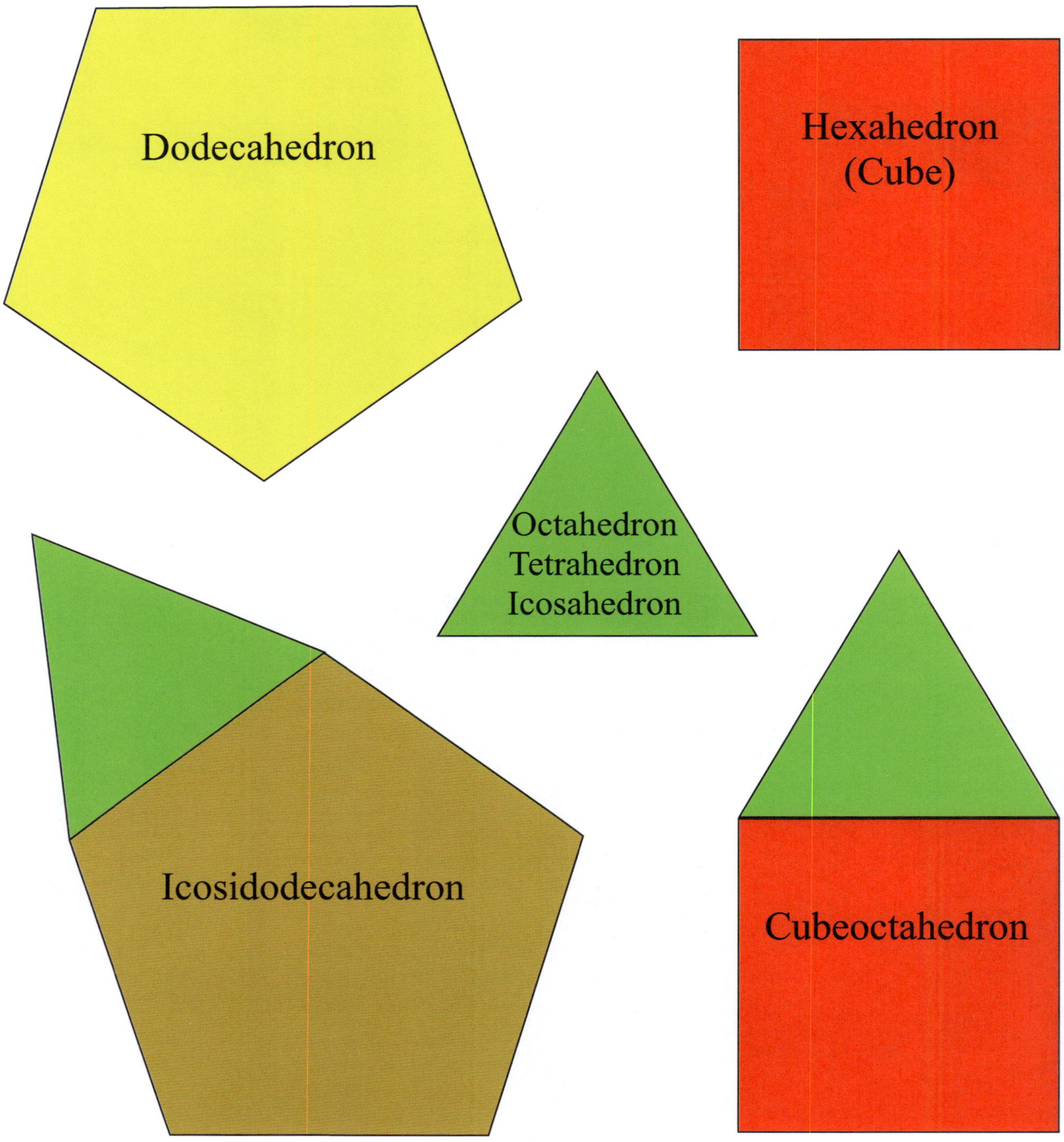

Tetrahedron

The Tetrahedron is nature's strongest basic structure. It is four triangles put together in their simplest configuration. To be classified as a Platonic Solid, all of the Faces, Edges and Corners (Vertexes) must be equal.

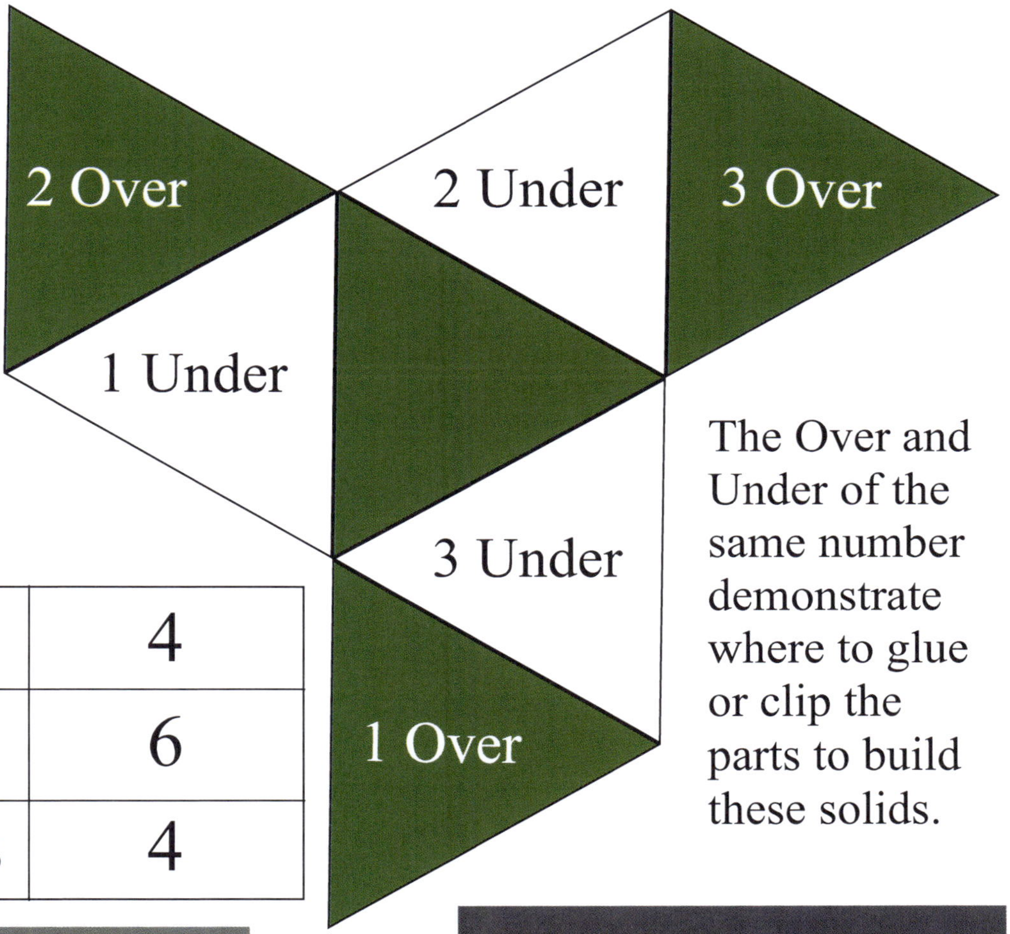

The Over and Under of the same number demonstrate where to glue or clip the parts to build these solids.

Faces	4
Edges	6
Vertices	4

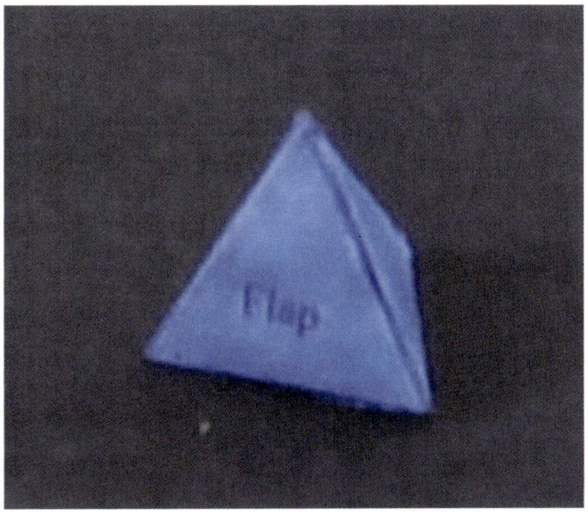

The previous page demonstrated the Pythagorean Theorem. These pages prove it, that is they are good for all right triangles.

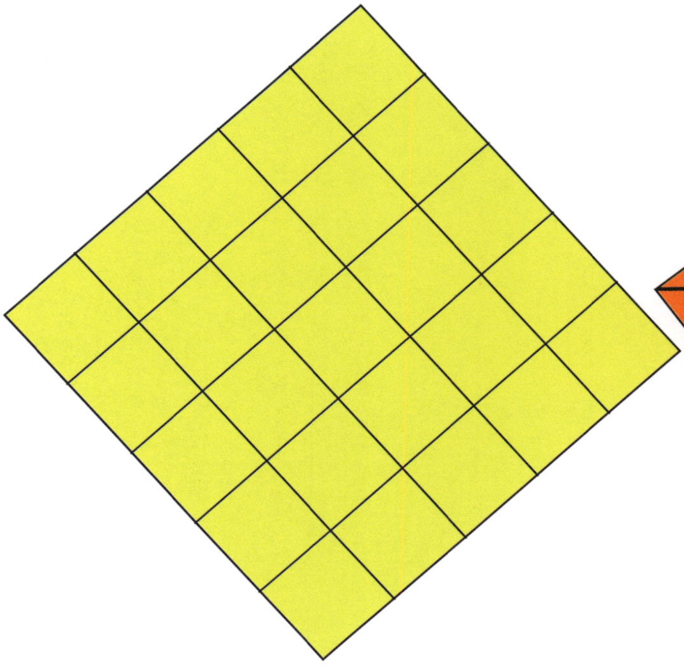

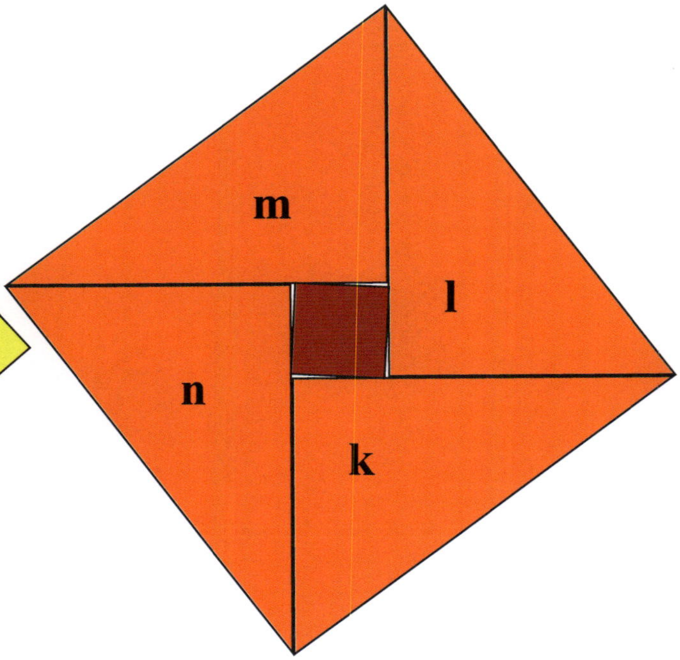

For any and all right triangles the square of the hypotenuse may be reproduced by placing four of the triangles together with the hypotenuse facing outward as shown.

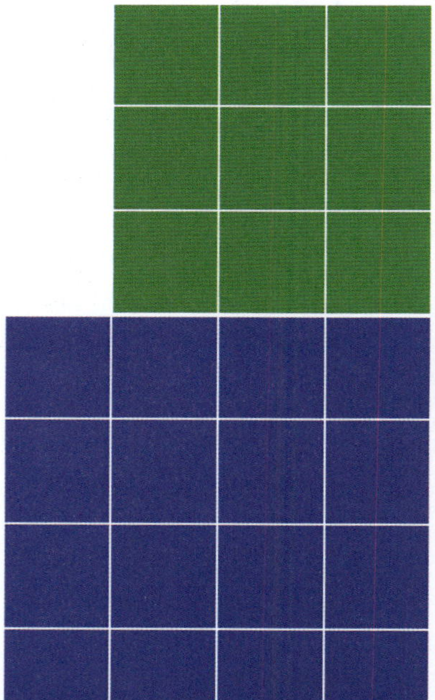

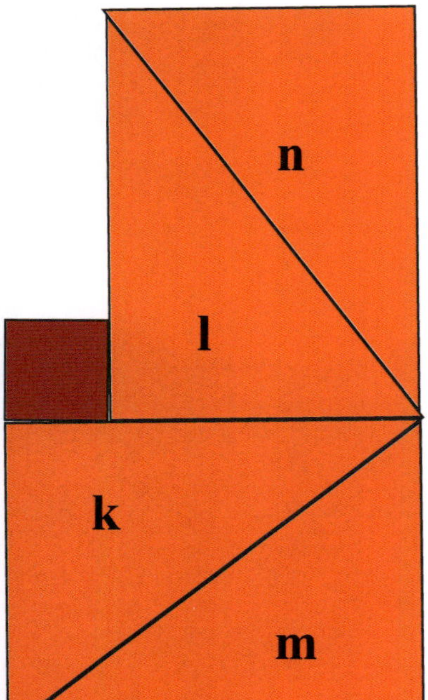

By moving two of the triangles a shape is created that is equal in area to the squares of the other two sides. This works for all right triangles.

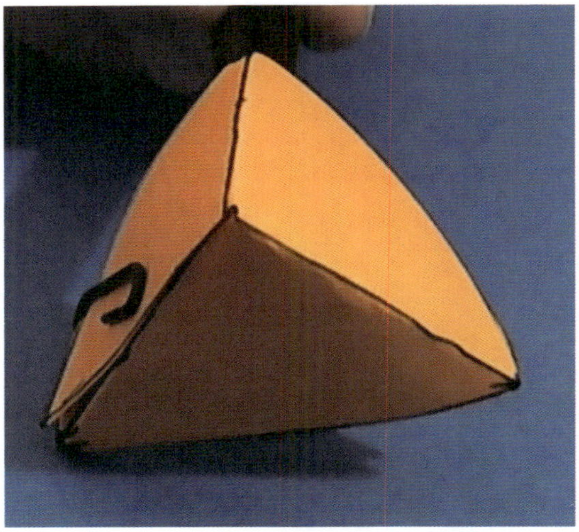 

The Tetrahedron has four equal faces, four equal corners, and six equal edges. Each of the Platonic Solids also has a partner solid known as its <u>Dual</u> that is created by connecting the center point of each face. The Tetrahedron is its own <u>Dual</u>. If you connect the center point of each face of the Tetrahedron you get another Tetrahedron. Each Solid also has what is called a <u>Rectification</u>. A <u>Rectification</u> is created by connecting the center points of each edge, or cutting the corners off all the way to the middle. The <u>Rectification</u> of the Tetrahedron is the Octahedron.

Octahedron

The Octahedron introduces the square. Although the faces are all triangles they come together in vertices of four. This gives the appearance of two four sided pyramids coming together at their square bases.

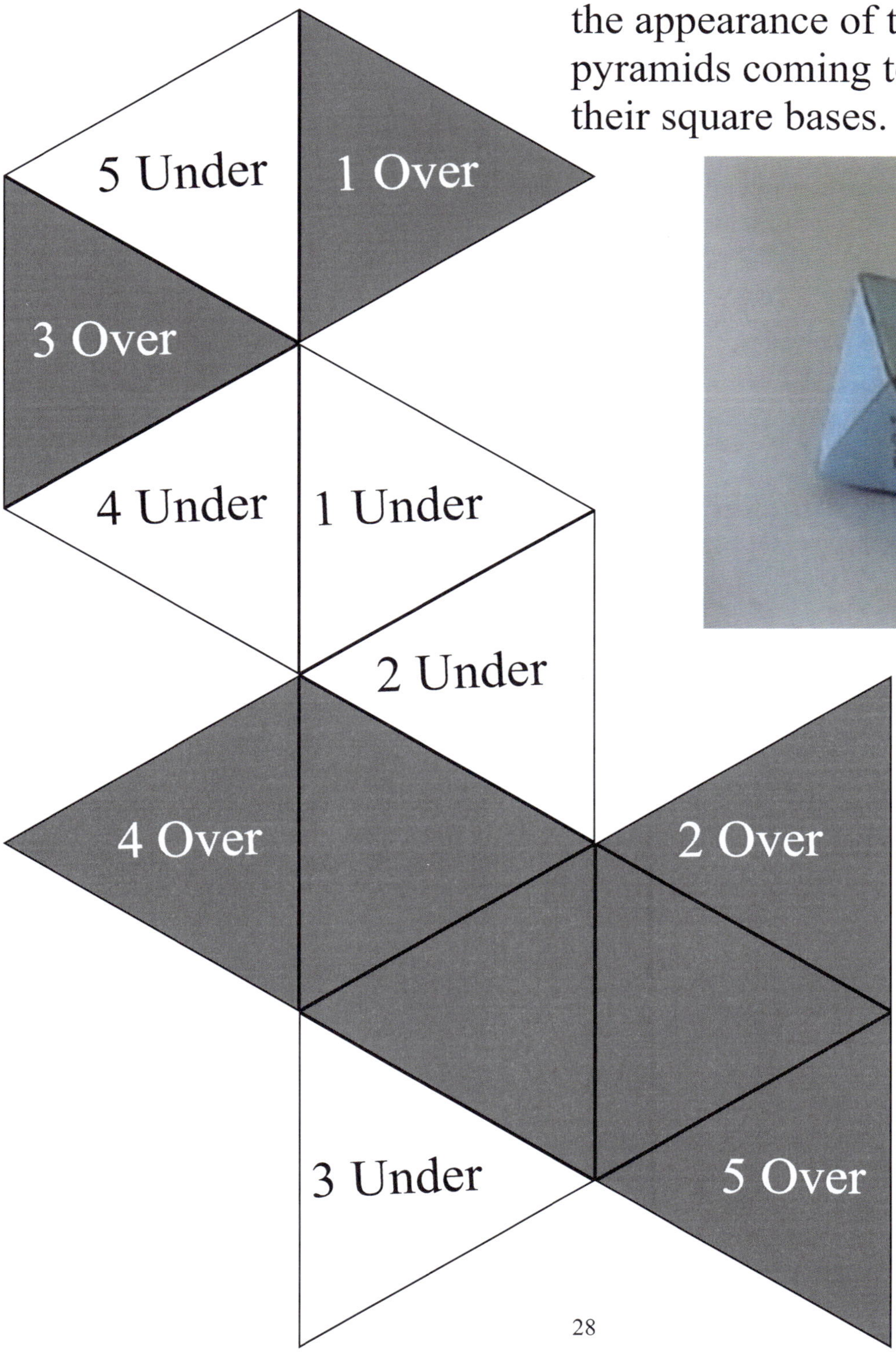

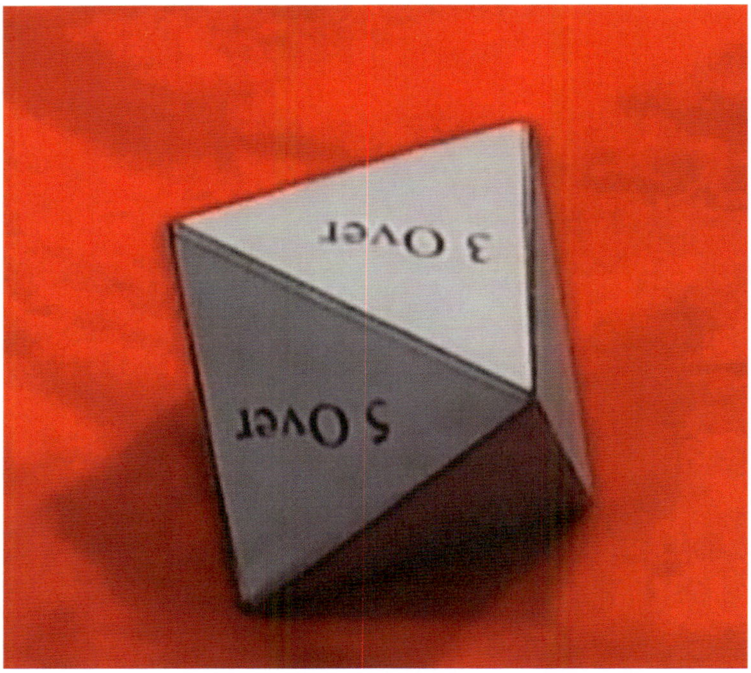

On second look, the connecting squares of the two pyramids actually recreate two other four sided pyramids when looked at from a different perspective.

Faces	8
Edges	12
Vertices	6

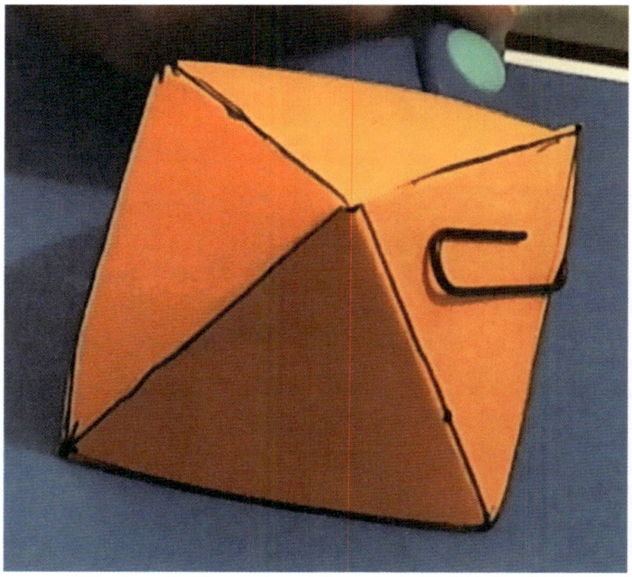

The Dual of the Octahedron is the Cube. If you connect the center points of the faces of the Octahedron you get a cube. A hint is that the Octahedron has eight faces and the Cube has eight corners. The rectification of the Octahedron is the Cube-Octahedron. If you connect the center points of the edges, or cut off the vertices all the way back to the center you get a Cube-Octahedron.

Hexahedron (Cube)

The Cube, our basic form for the measurement of volume, is dominated by 90° angles. We use Cubic Inches, Cubic Feet, Cubic Meters, to measure 3-Dimensional Space.

The Dual of the Cube is the Octahedron. If you connect the center point of the faces of the Cube you get an Octahedron. Its rectification is the Cube-Octahedron.

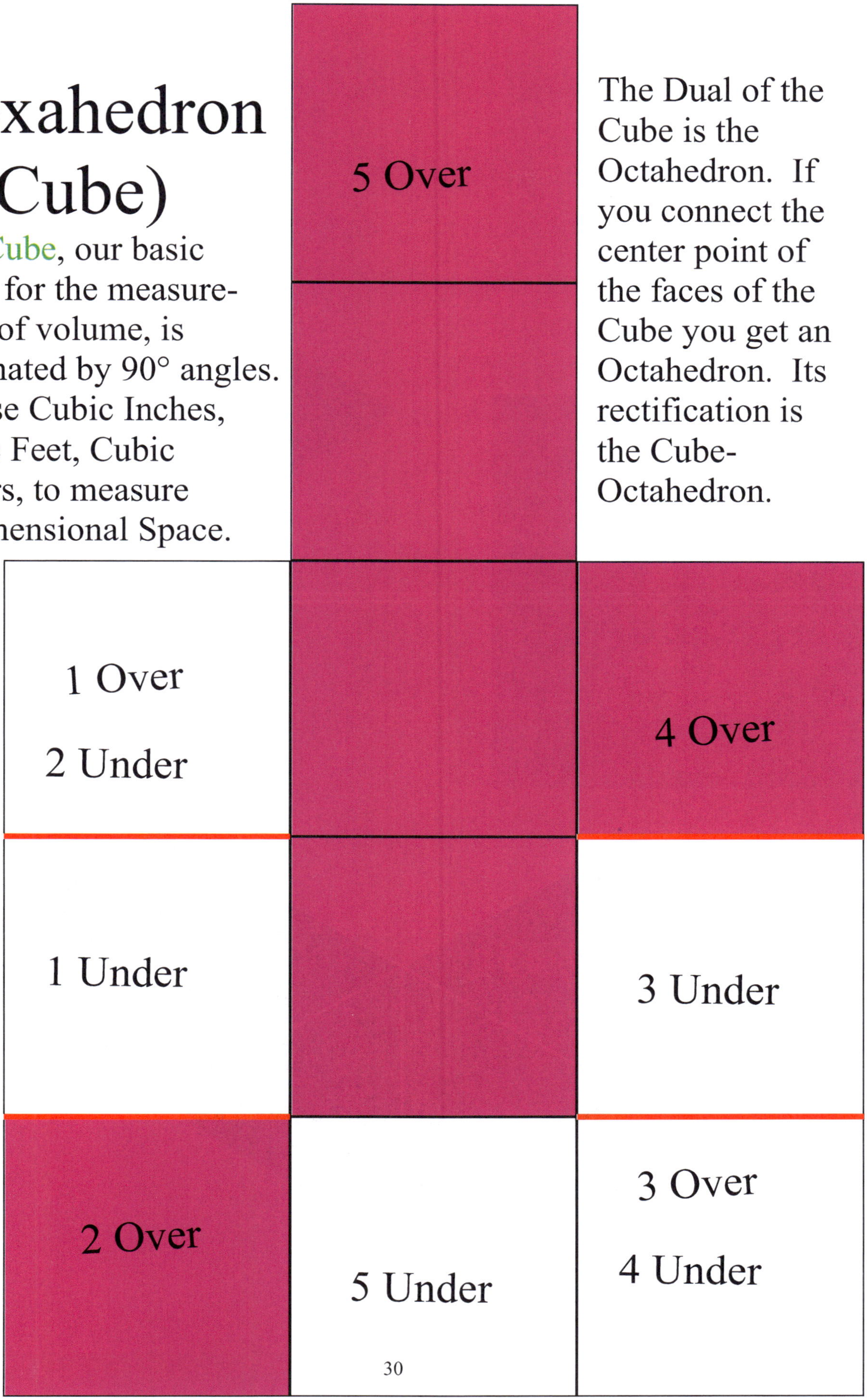

Faces	6
Edges	12
Vertices	8

The Cube is created using the square pattern we derived from drawing the circle and the inscribed hexagon.

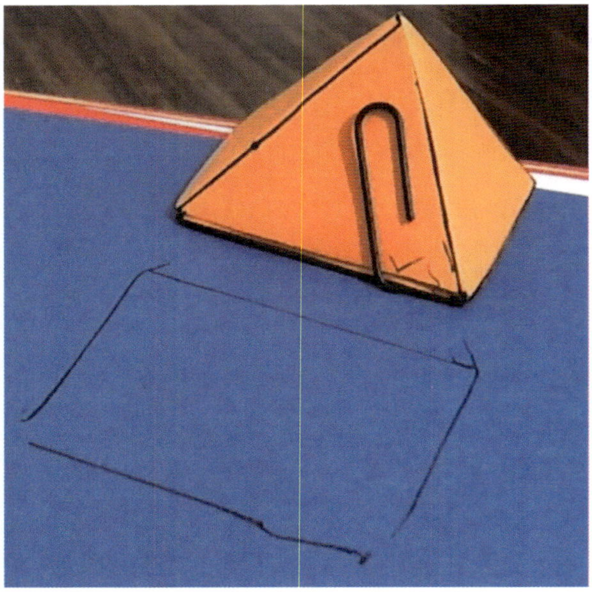

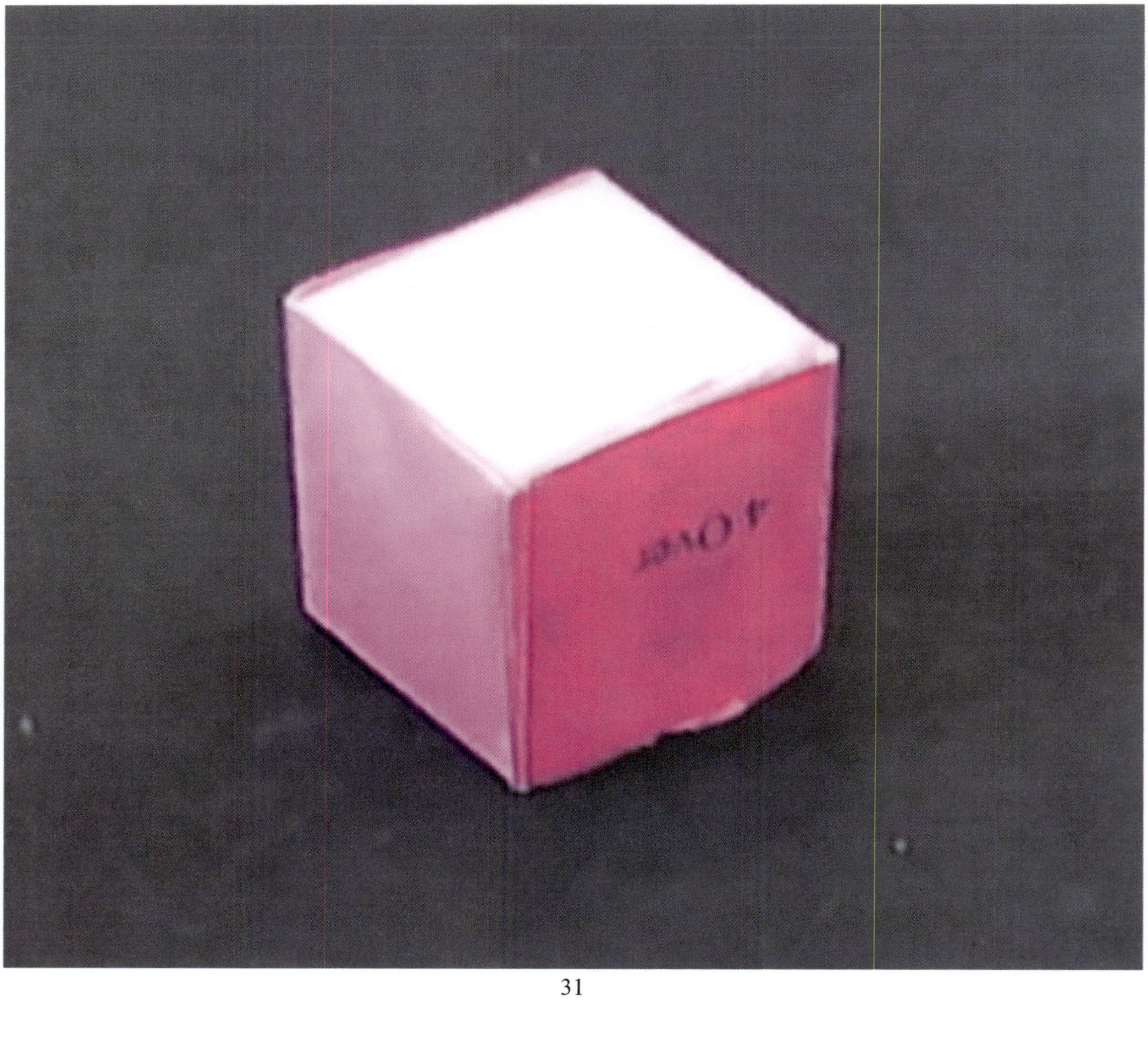

Icosahedron

The Icosahedron is the largest of the Platonic Solids that have triangular faces. Five triangles come together at each vertex and the base of these five triangles form the pentagon. This is the same pentagonal pyramid that we saw by folding under one of the triangles of our hexagon. The pentagon is also the first appearance of the Golden Ratio, ϕ.

The Dual of the Icosahedron is the Dodecahedron. It has 20 faces and the Dodecahedron 20 vertexes.

Faces	20
Edges	30
Vertices	12

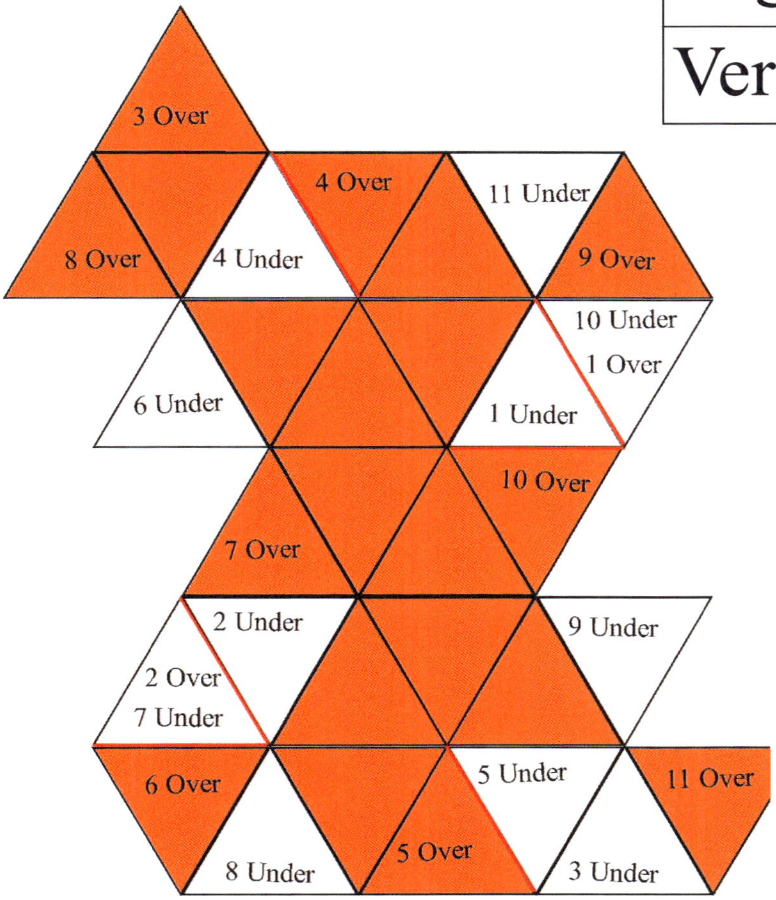

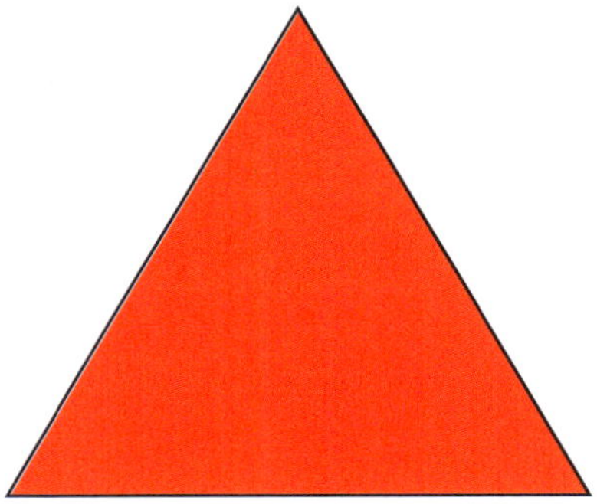

The Rectification of the Icosahedron is the Icosidodecahedron.

The Icosahedron makes a 20 face sphere. Each Vertex has 5 triangles coming together. It introduces the pentagon, which introduces the Golden Ratio.

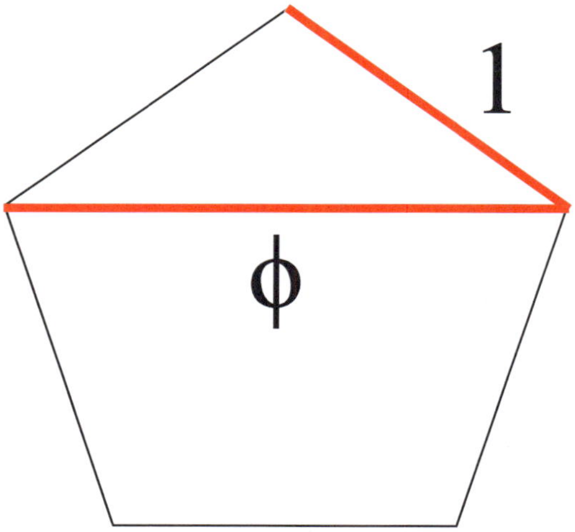

In the Pentagon, if the edge is 1 then the cross beam is Phi.

This is the same configuration as the five sided triangle configuration from our hexagon.

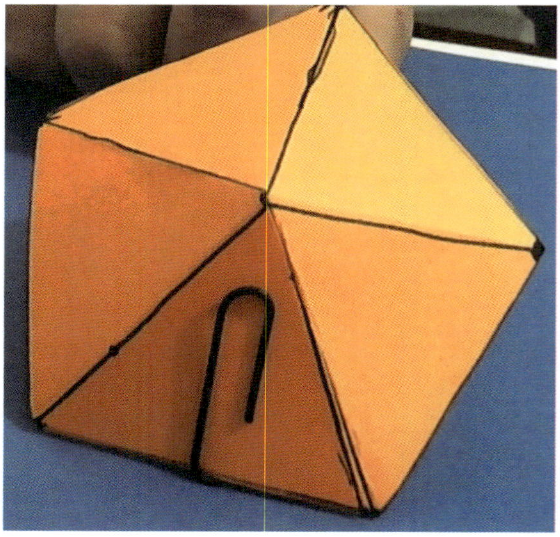

Dodecahedron

The Dodecahedron is the last of the Platonic Solids. It is formed by putting twelve pentagons together. The Vertices are formed by the intersection of three pentagons.

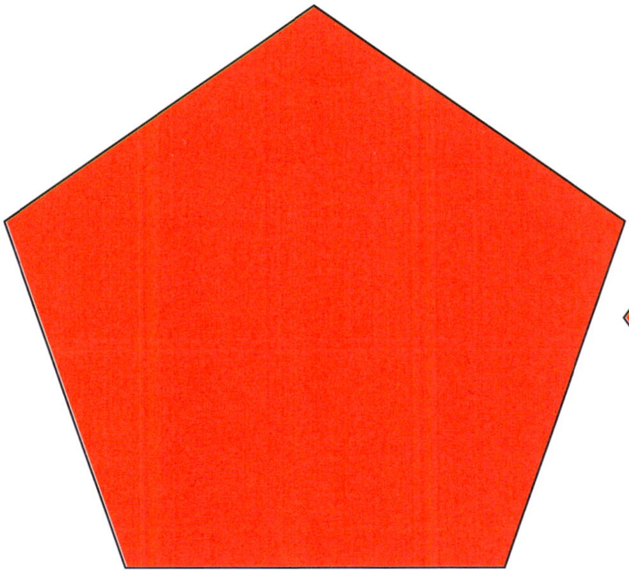

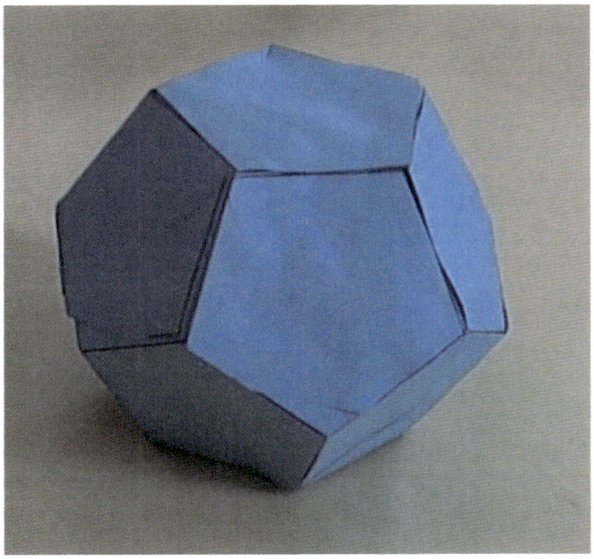

Faces	12
Edges	30
Vertices	20

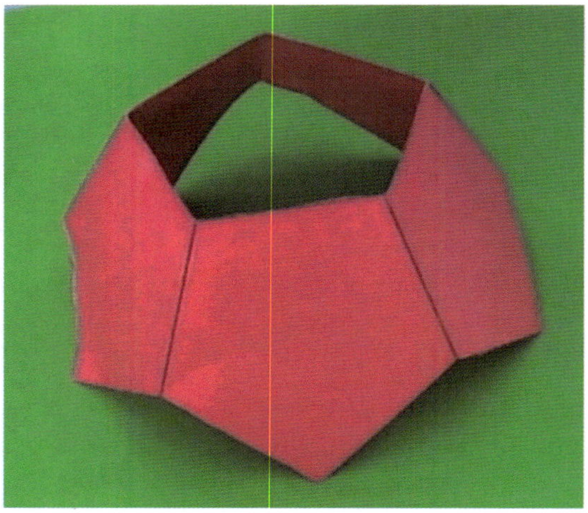

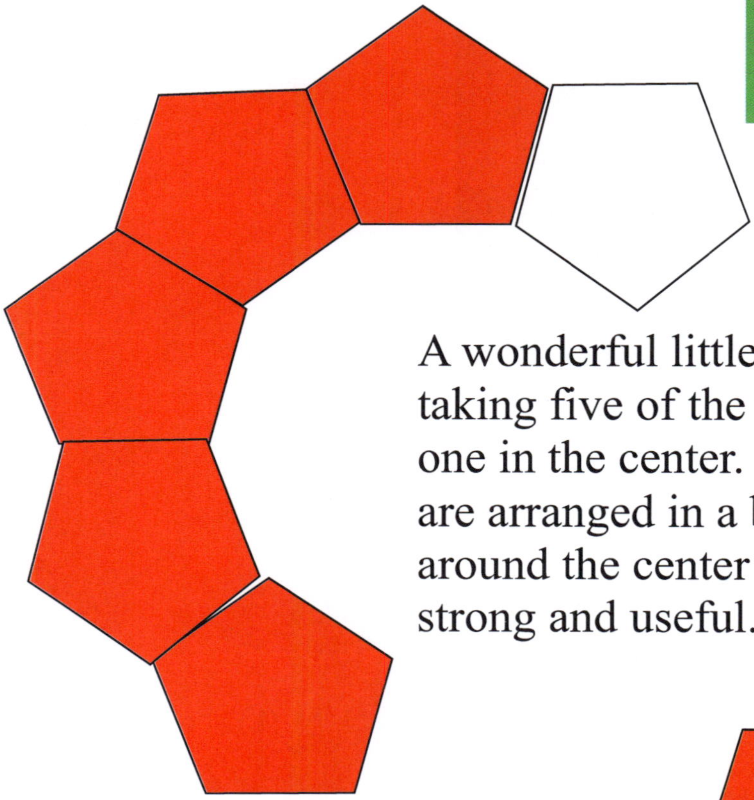

A wonderful little basket can be made by taking five of the pentagons arrayed around one in the center. Another six pentagons are arranged in a belt that is wrapped around the center group. The basket is strong and useful.

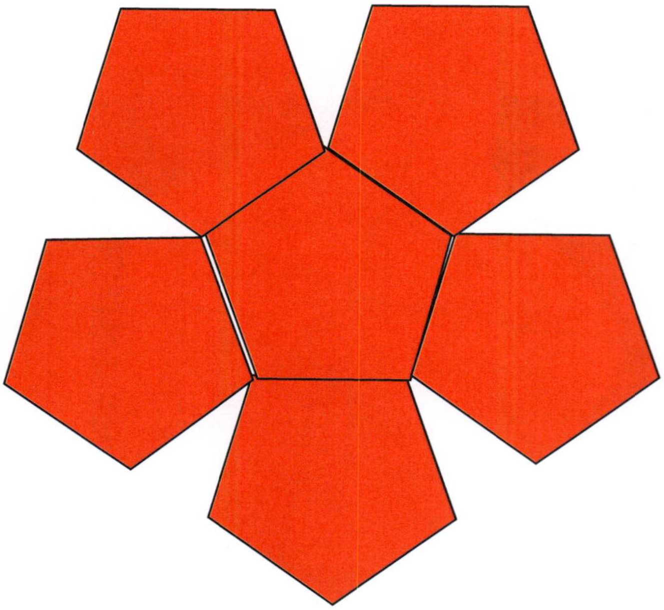

The Dual of the Dodecahedron is the Icosahedron. Notice the Dodecahedron has 12 Faces and the Icosahedron has 12 Vertices.

Its Rectification is the Icosidodecahedron.

Cube-octahedron

The Cube-octahedron is a combination of the Cube and the Octahedron. It is called a rectification of either of these shapes. There are several ways to form this shape. One is to put six squares and eight triangles together, with each triangle always next to a square and vice versa.

Faces	14
Triangles	8
Squares	6
Edges	24
Vertices	12

The second way to make this shape is to shave the corners off of the cube. Continue shaving the corner until you reach the middle of the side. This is the rectification of the Cube.

The third way to make this shape is to shave the corners off of the Octahedron. Continue shaving the corner until you reach the middle of the side. This is the rectification of the Octahedron.

Buckminster Fuller called the Cube-octahedron, Vector Equilibrium. It is the only solid in which the edges and the internal struts are all exactly the same length. What keeps this from being a Platonic Solid is that the faces are not all the same. Some are squares and some are triangles.

We will see this more clearly when we get to the second part of the book and look at the internal structures of the Platonic Solids.

We will recognize that the triangles we see here are connected internally to three other triangles of equal size and the squares that we see are connected to four triangles in the same configuration that we derived from folding triangles under with the hexagon.

Icosidodecahedron

The Icosidodecahedron is a combination of the Icosahedron and the Dodecahedron. It is called a rectification of either of these shapes. There are several ways to form this shape. One is to put twelve pentagons and twenty triangles together, with each triangle always next to a pentagon and vice versa.

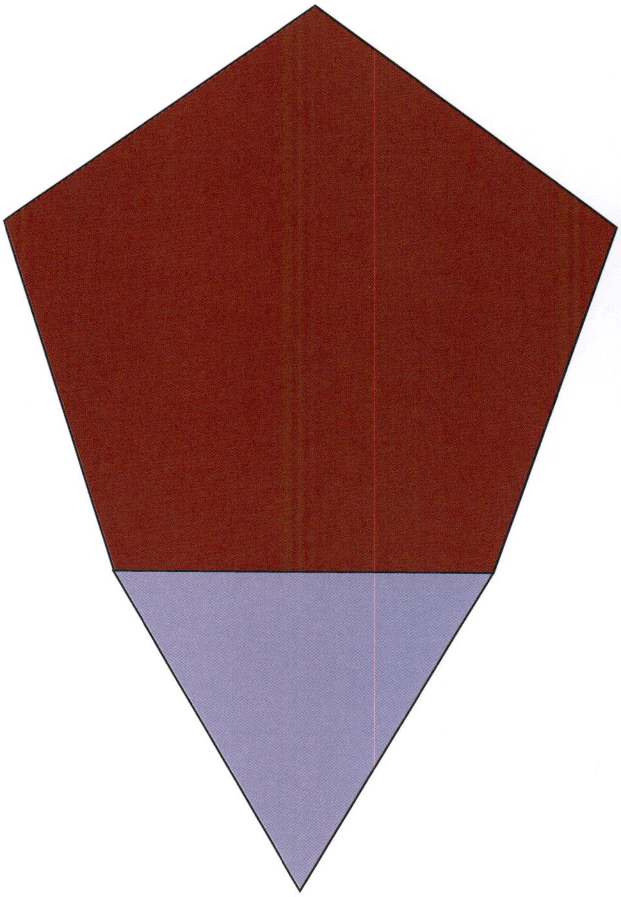

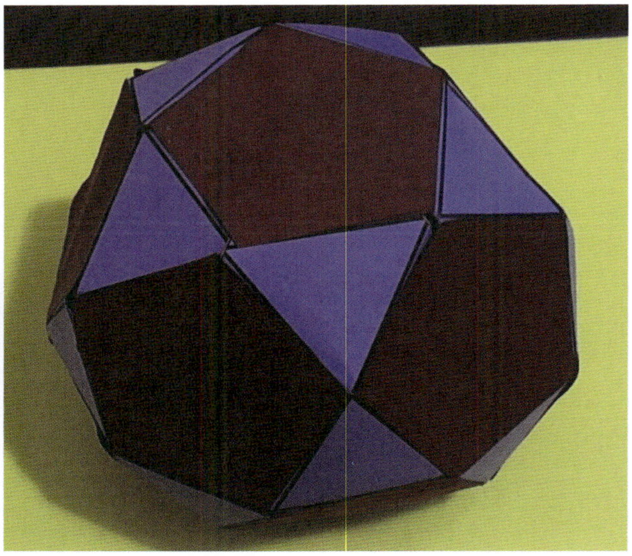

Faces	32
Triangles	20
Pentagons	12
Edges	60
Vertices	30

Rumored to be one of Leonardo Da Vinci's favorite shapes.

Icosidodecahedron

This shape has captured the imagination of countless mathematicians, artists, and thinkers throughout history. Its sheer beauty is one reason as well as the fact that it can be built by using six large circles. The other may be its relation to the Golden Ratio or as Kepler called it the Extreme and Mean Ratio. It is chock full of it. It has a pentagonal face which has Phi in it, but it is the interior that is the most intriguing. If the edge of this shape is one, the Radius is Phi, 1.6182...

The second way to make this shape is to shave the corners off the Icosahedron Continue shaving the corner until you reach the middle of the side. This is the rectification of the Icosahedron.

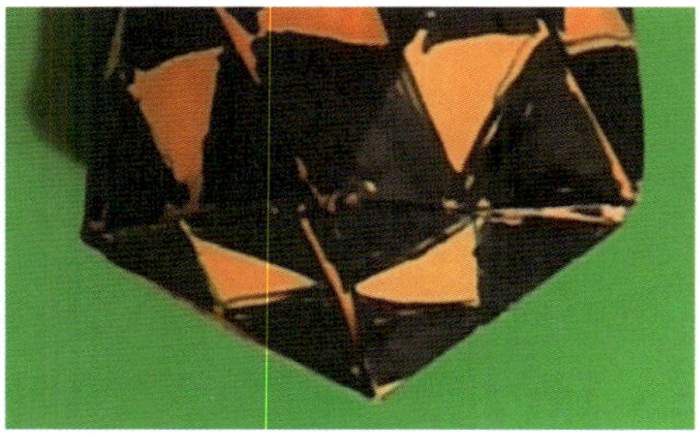

The third way to make this shape is to shave the corners off the Dodecahedron. Continue shaving the corner until you reach the middle of the side. This is the rectification of the Dodecahedron.

The fourth way to make this shape is to weave together six circles as we do with colored foam strips. These circles are clear in the Hoberman Sphere and in Roger von Oech"s X-Ball, both available commercially. Or seek out Leonardo Da Vinci's version.

Hoberman Sphere. Buy it!

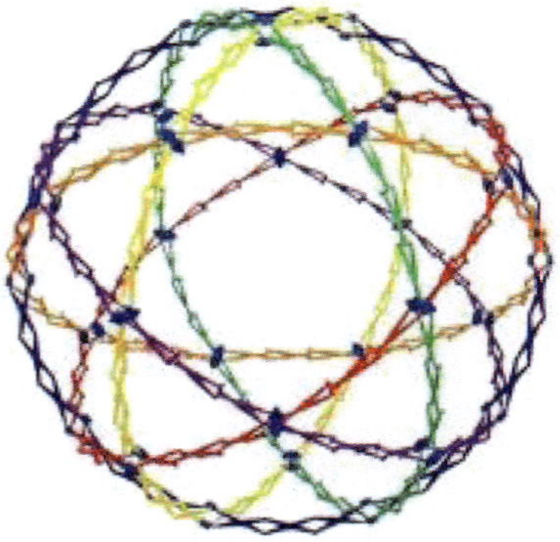

Leonardo Da Vinci's Rendering. Find it!

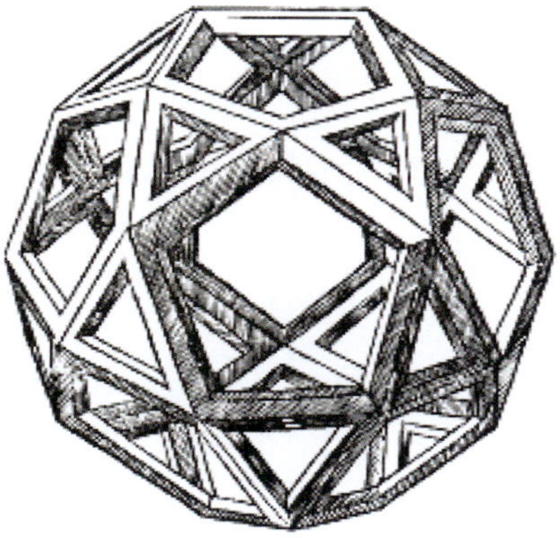

Roger von Oech's X-Ball. Buy it!

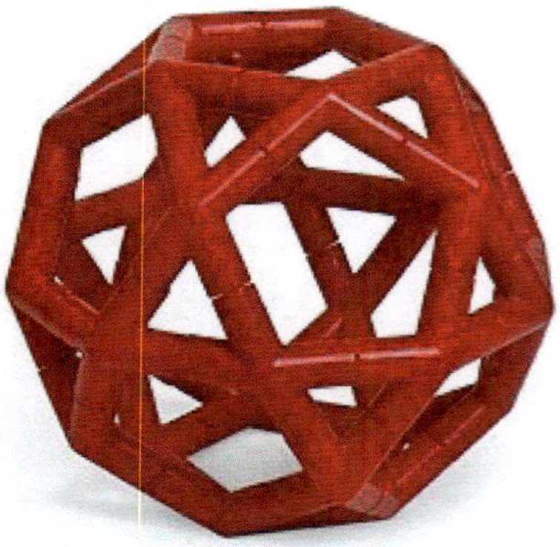

Number and Measure

I. <u>The Number One</u>

One of the great problems of mathematics is deciding where to start. All of the ideas in math are related to each other and reinforce each other. In this book we have decided to start with the **number one, 1**. The number one itself is filled with subtle distinctions, even mysteries.

> ***One can be a single thing.***
>
> ***One can be a combination of things.***
>
> ***One can be everything.***

When we ***measure*** things, **one** is the thing to which we refer. In our little book **one, 1** can refer to different measurements. We measure length, area and volume.

1 inch

Our measure for length is the English Inch. →

Our measure for area is the Square Inch. **1 in²** →
This process is called squaring. $\mathbf{1 \times 1 = 1 = 1^2}$

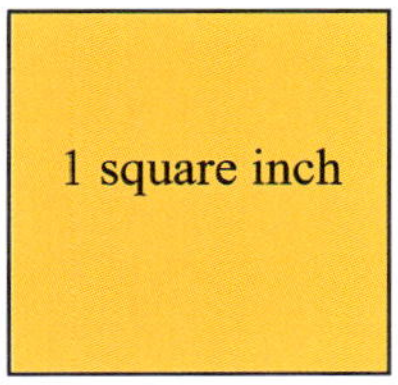

Our measure for volume is the Cubic Inch. **1 in³**
This process is called cubing. $\mathbf{1 \times 1 \times 1 = 1 = 1^3}$

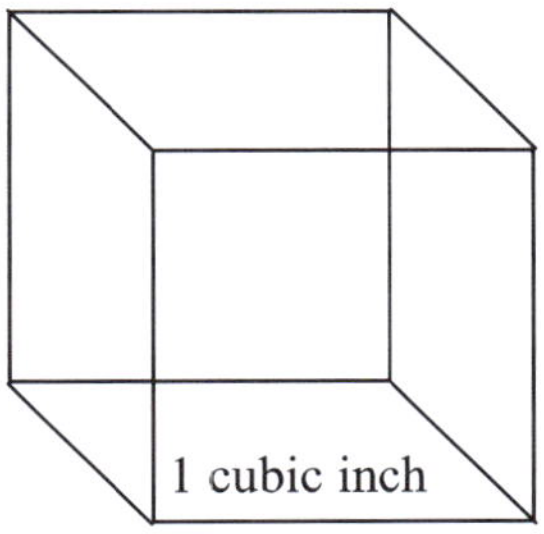

One, the number, doesn't have to measure anything. It can be just a number. This is called a Dimensionless Number. Numbers can be completely abstract. They can exist in our imagination. At times, one may not refer to anything at all other than the number itself. Even in this case it still has rules and patterns even though it does not refer to anything in the real-world.

And sometimes One is just an idea of unity, a la John Lennon, "and the World would live as One."

The Incredible Effectiveness of Mathematics

Length, area and volume are not the only dimensions that we measure with numbers. Temperature and angles are measured in degrees. Weight and mass are measured in pounds and grams. Liquids are measured in gallons and liters. But the most amazing thing about mathematics is that it can often switch between dimensions and dimensionless numbers. Squaring a number gives you a measure in square inches, but it can also be, and often is, treated as dimensionless.

Whole Numbers

All the numbers that are made of whole ones (1) are called whole numbers. These are the counting numbers. In this book we use the following whole numbers:

1, 2, 3, 4, 5, 7, 9, 12, 16, 25, and 49.

Fractions and Ratios

Parts of whole numbers are called fractions. All whole numbers and fractions can be stated as a Ratio. A Ratio is when we compare one number to another. We show a ratio by placing a line or a colon between the two numbers like **1/3** or **1:3.** It is just like division. 1/3 is the same as 1 divided by 3. All numbers can be listed as a ratio. The number 5 can be stated as the ratio 5/1. The fraction one half can be stated as the ratio 1/2.

Rational Numbers

When both the top (numerator), first, and bottom (denominator), second number of a ratio are whole numbers, that number is called a rational number.

Irrational Numbers

When a number cannot be stated by a ratio of two whole numbers, then that number is called an irrational number. **(This does not mean the number is crazy)** We use symbols to designate these numbers. In this book we will look at the irrational numbers $\sqrt{2}$, $\sqrt{3}$, $\sqrt{5}$, π , **e,** and **Φ.** Even though these numbers cannot be stated as a whole number ratio, they are critically important to many processes in math. **If they seem Greek to you, they are.**

We've come full circle, but keep in mind where we ended up and see where we go when we venture into the world of **Squares.**

The Squares

We operate in a square world.
The foundation of that world is the
Unit Square. Notice: $1^2 = 1 = \sqrt{1}$
1 is called the Multiplicative Identity

$1^2 = 1$
$2^2 = 1 + 3 = 4$
$3^2 = 1 + 3 + 5 = 9$
$4^2 = 1 + 3 + 5 + 7 = 16$
$5^2 = 1 + 3 + 5 + 7 + 9 = 25$

See how the squares grow.
2^2 is 3 more than 1^2.
3^2 is 5 more than 2^2.
4^2 is 7 more than 3^2.
5^2 is 9 more than 4^2. etc.

$1^2 = 1x1 = 1$

$2^2 = 2x2 = 4$

$3^2 = 3x3 = 9$

$4^2 = 4x4 = 16$

$5^2 = 5x5 = 25$

Squaring means multiplying a number by itself.

The number being squared can be thought of as the side of a square. The result is the number of 1x1 squares inside the larger square. You can also work backwards. If you know the number of 1x1 squares in the larger square, you can find the side. This is known as the square root.

This allows us to locate all the other square roots between the perfect squares.

$1 < \underline{\sqrt{2}, \sqrt{3}} < 2$

$2 < \underline{\sqrt{5}, \sqrt{6}, \sqrt{7}, \sqrt{8}} < 3$

$3 < \underline{\sqrt{10} \rightarrow \sqrt{15}} < 4$

$4 < \underline{\sqrt{17} \rightarrow \sqrt{24}} < 5$

$1 = \sqrt{1}$

$2 = \sqrt{4}$

$3 = \sqrt{9}$

$4 = \sqrt{16}$

$5 = \sqrt{25}$

The Pythagorean Theorem
The 3, 4, 5 Triangle
The Rope Stretchers Triangle

As Kepler has told us, one of the great treasures of mathematics is the Pythagorean Theorem. This theorem lies at the heart of much of mathematics and is at the center of such basic functions as squaring and taking square roots. Stated briefly, the Pythagorean theorem tells us that in right triangles the square of the hypotenuse is equal to the sum of the squares of the other two sides. The hypotenuse is the side opposite the right angle, the 90° angle.

$$a^2 + b^2 = c^2$$
$$3^2 + 4^2 = 5^2$$
$$9 + 16 = 25$$

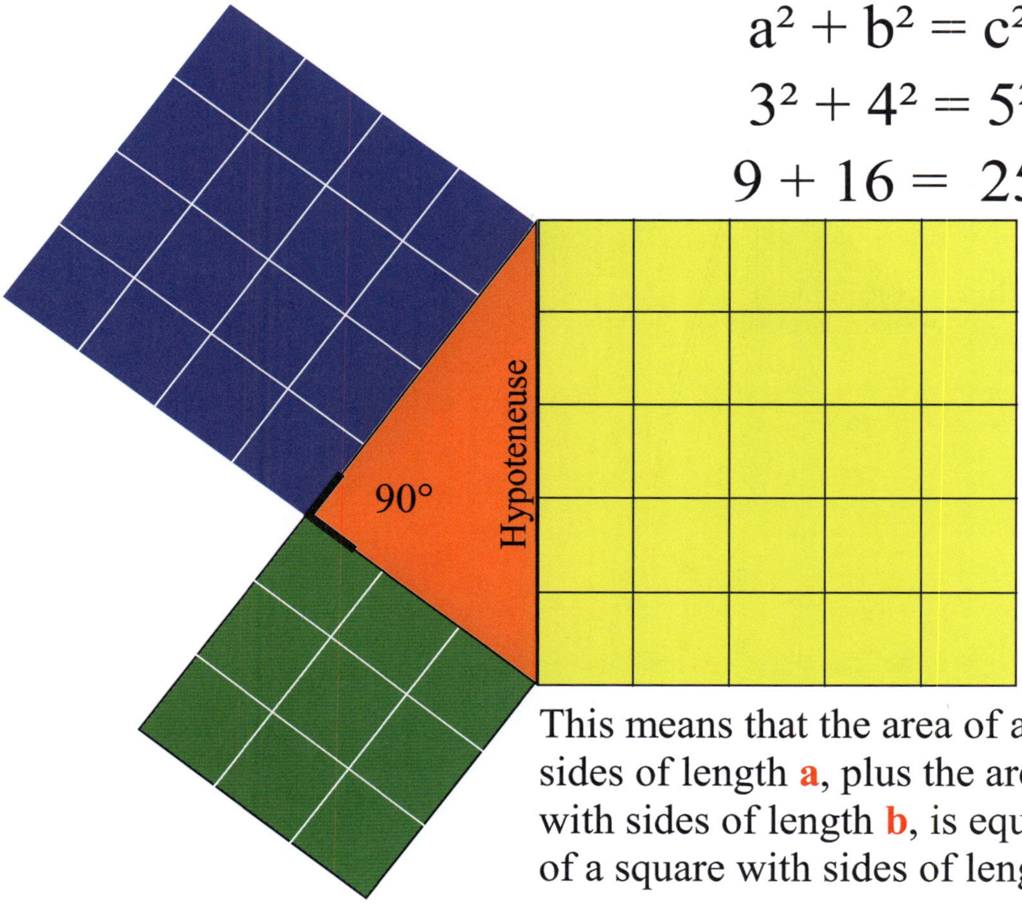

This means that the area of a square with sides of length **a**, plus the area of a square with sides of length **b**, is equal to the area of a square with sides of length **c**.

We get our 3,4,5 triangle by making a rectangle out of 3 rows of 4 unit squares or 4 rows of 3. We then cut it along the diagonal and we produce two 3.4.5 triangles.

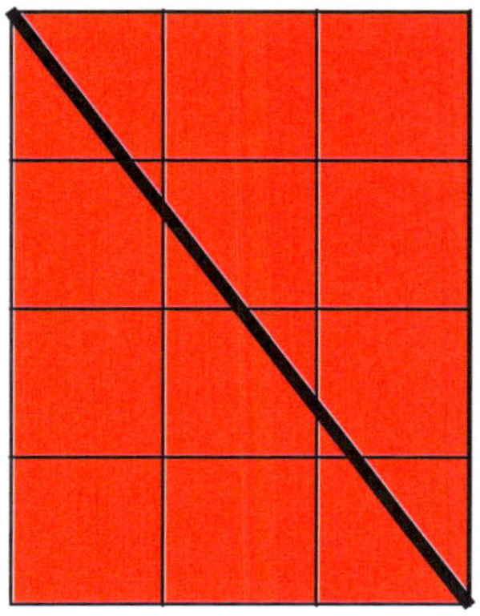

The angles of every triangle sum up to 180°.

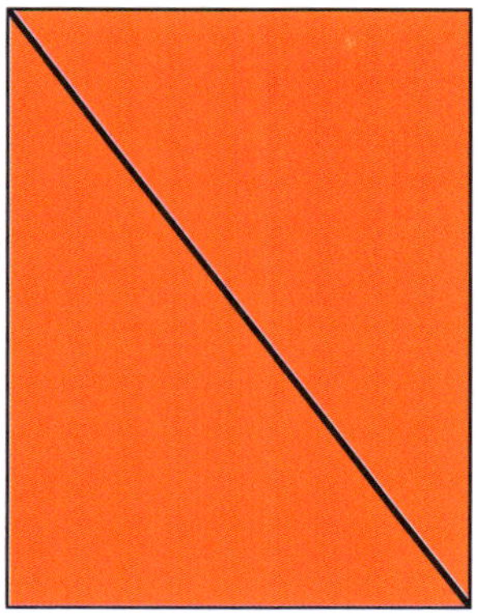

In all right triangles, in which one of the angles is 90°, the other two angles will sum up to 90°.

As we have seen, the two non right angles in a right triangle will sum to 90°, which allows us to create a 180° angle with these two angles and a 90° angle.

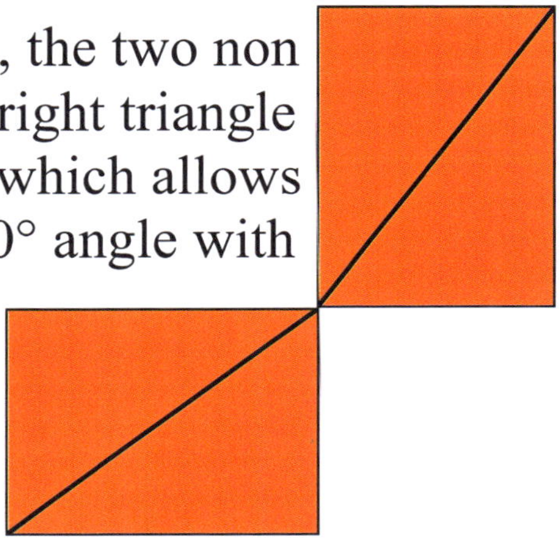

We can also put two of our rectangles together in a manner that allows us to have two 90° corners facing each other and leaving space for two more 90° angles.

We can also create a square by taking four right triangles and facing the hypotenuse outward. This square is equal to the square of the hypotenuse.

The Egyptians used a rope divided and marked into **12** equal lengths with a loop at each end. They looped the two ends around a pole. They took two other poles and stretched them tight at the **4** and **7** positions. This told them that the angle at the **4** position was a right triangle, an angle of 90°.

The Pythagoreans valued this triangle because all the sides were whole numbers. That was not always the case. One triangle in particular gave them fits. That was the isosceles right triangle. What if the sides adjacent to the right angle were equal and both of length 1. That meant their squares added up to 2. What number multiplied by itself is equal to 2? We will soon see.

Before you go. Take the 1 x 1 blocks and fill up the 3 x 3 and the 4 x 4 squares. Once you have done this, move the 1 x 1 squares to the 5 x 5 square. This will fill that square exactly.

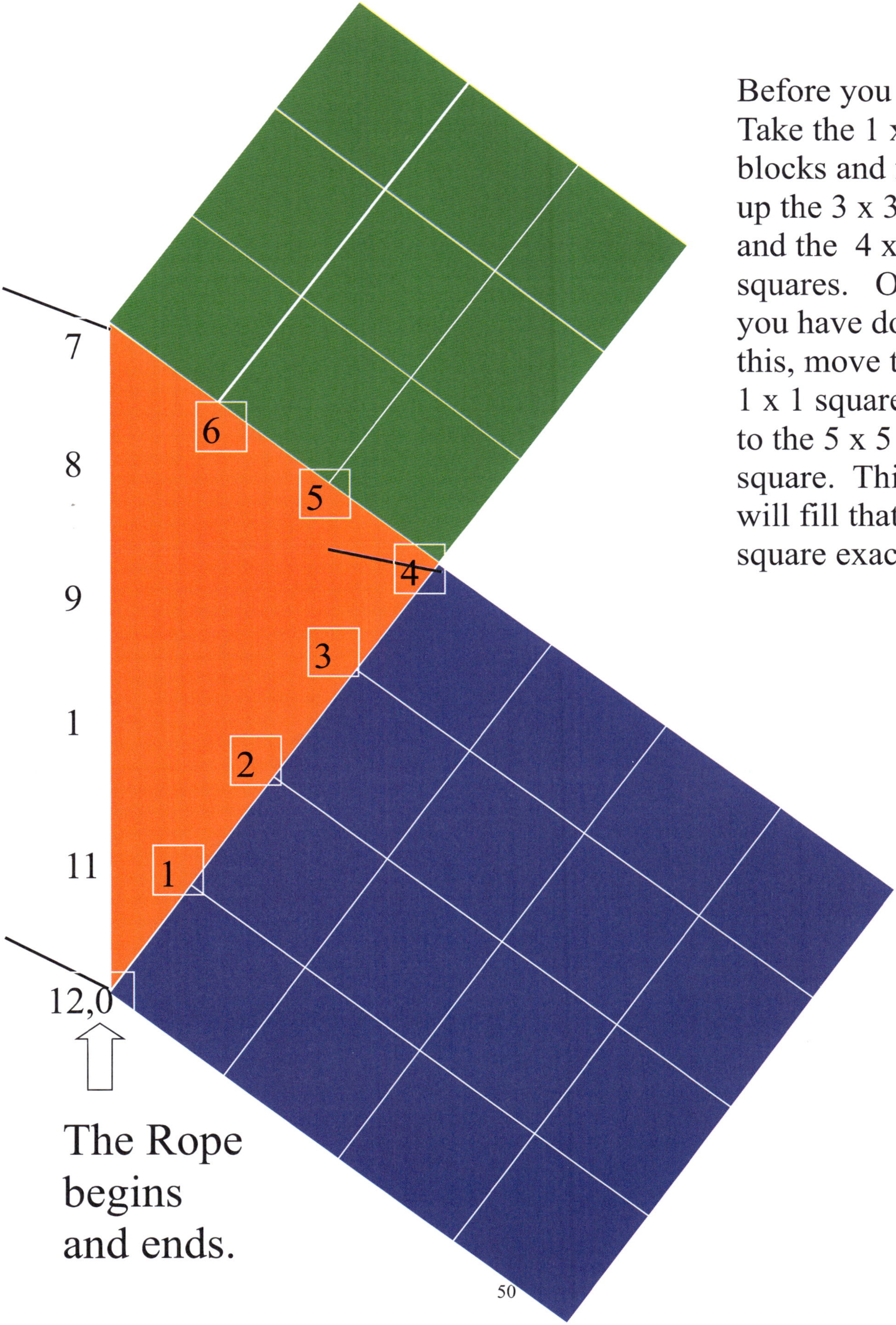

The Rope begins and ends.

Another universal proof of the Pythagorean Theorem may be achieved by placing four of the triangles around the square of the hypotenuse, as shown. This gives a square with sides equal to the sum of the lengths of the other two sides.

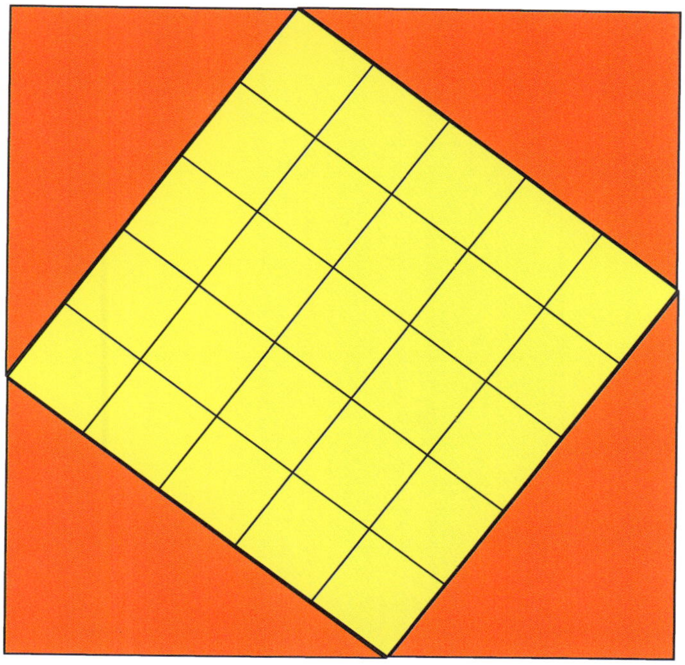

Another square is created by placing the squares of the other two sides in diagonal relation to each other and four of the triangles arranged in two rectangles that would fill the square out completely. This square has length of sides equal to the sum of the sides of the other two squares.

If we then grant that these two squares are of equal size we may then remove one triangle from each of them and the shapes should maintain their equality of area. We would then remove another and another and another until the triangles are all gone. What is left on the top is the square of the hypotenuse and on the bottom the squares of the other two sides.

The Pythagorean Theorem

1, 1, √2 Triangle, 45°, 45°, 90°

Isosceles Right Triangle

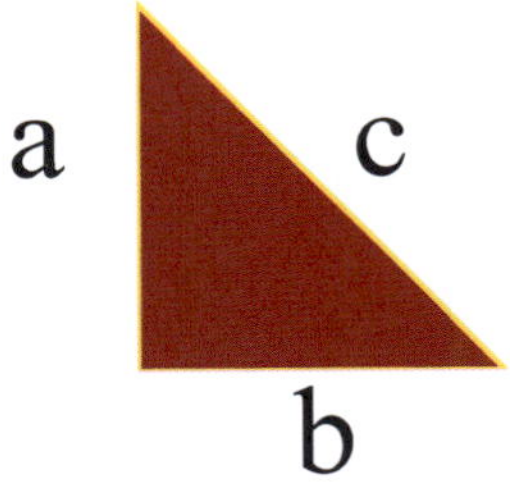

$$a^2 + b^2 = c^2$$
$$1^2 + 1^2 = c^2$$
$$1 + 1 = 2 = c^2$$
$$c = \sqrt{2}$$
$$c = 1.414\ldots$$

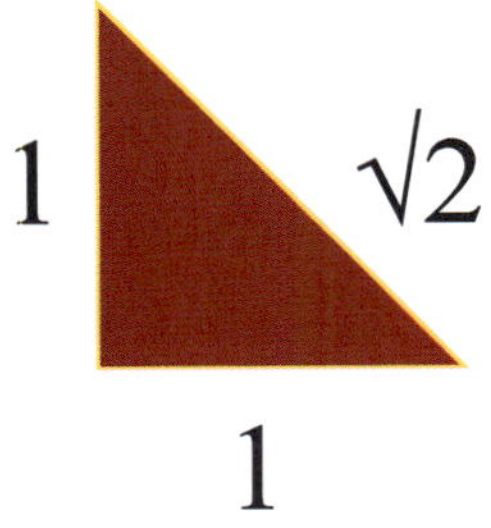

This triangle has a hypotenuse that cannot be expressed in whole numbers. It is the door to a new concept, the square root.

A square root asks the question: what is the length of the sides of a square that will give a certain area. A square must have sides of 1.414.. to give an area of 2.

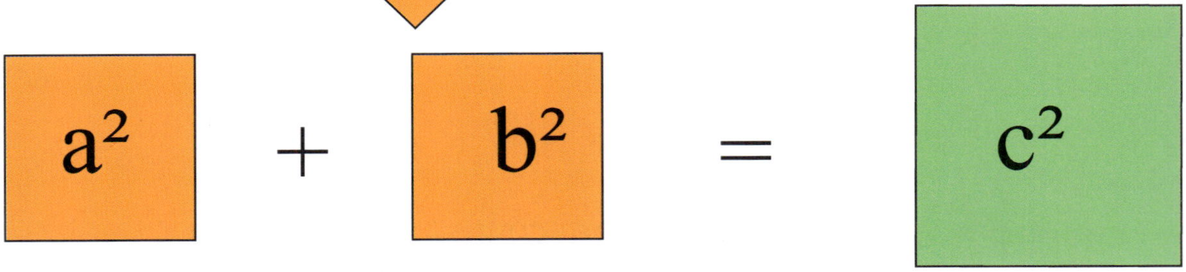

This triangle also has a beautiful demonstration.

We can also use both of our universal proofs on this triangle. The first one looks a little different than the 3,4,5 triangle because when we put four of the triangles together with the hypotenuse facing out, there is no little square in the middle. This is because the two sides are the same length and no square is created. Otherwise it is the same.

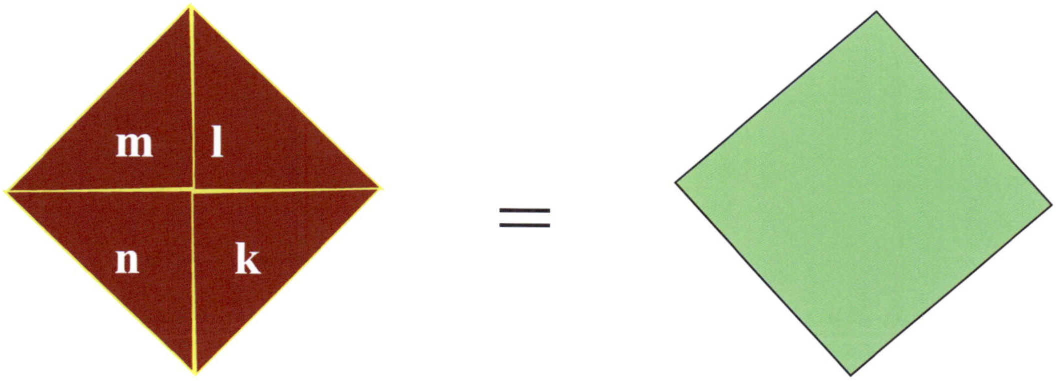

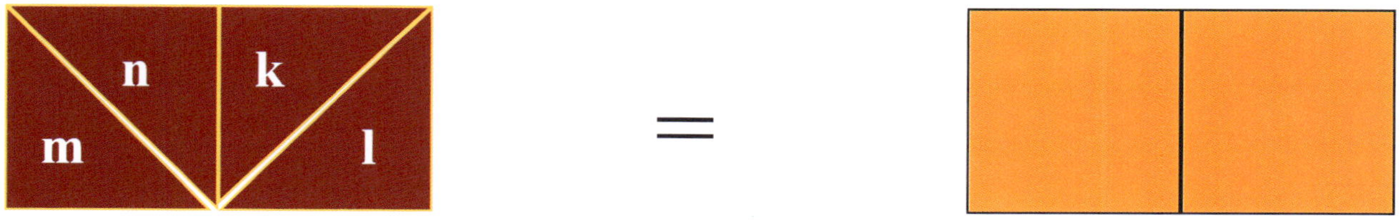

The other proof is the same. We create two squares whose sides are equal. We then remove the triangles one at a time and we see that the square of the hypotenuse is equal to the sum of the squares of the other two sides.

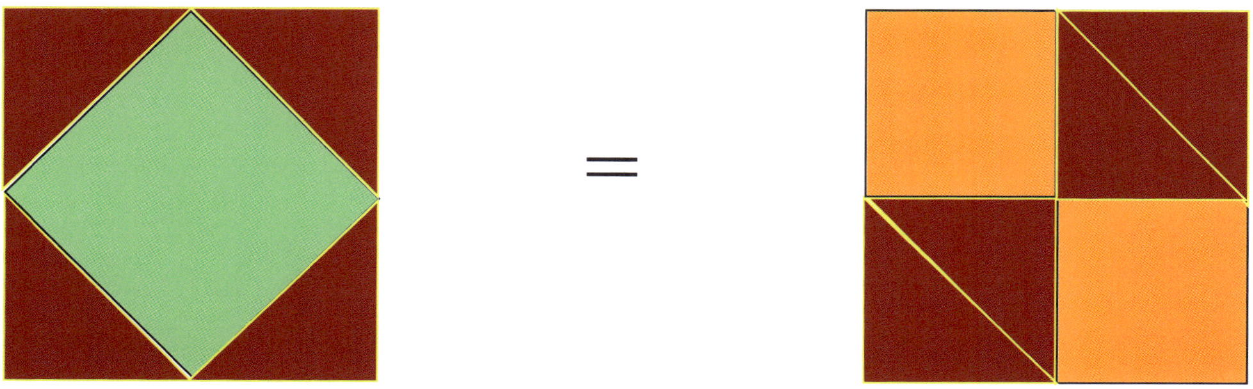

The Spiral of Theodorus

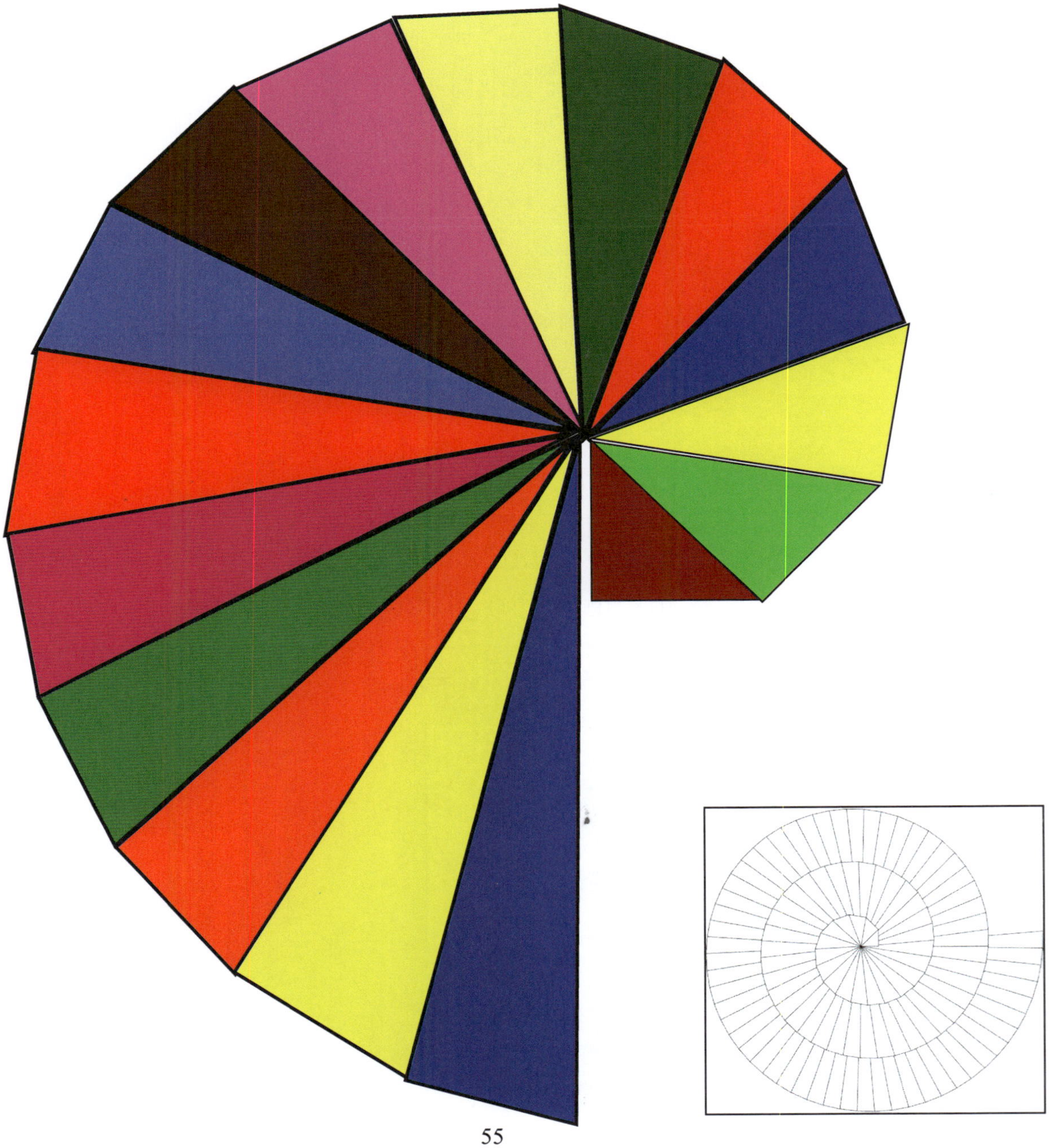

The First 8 Triangles of the Spiral of Theodorus

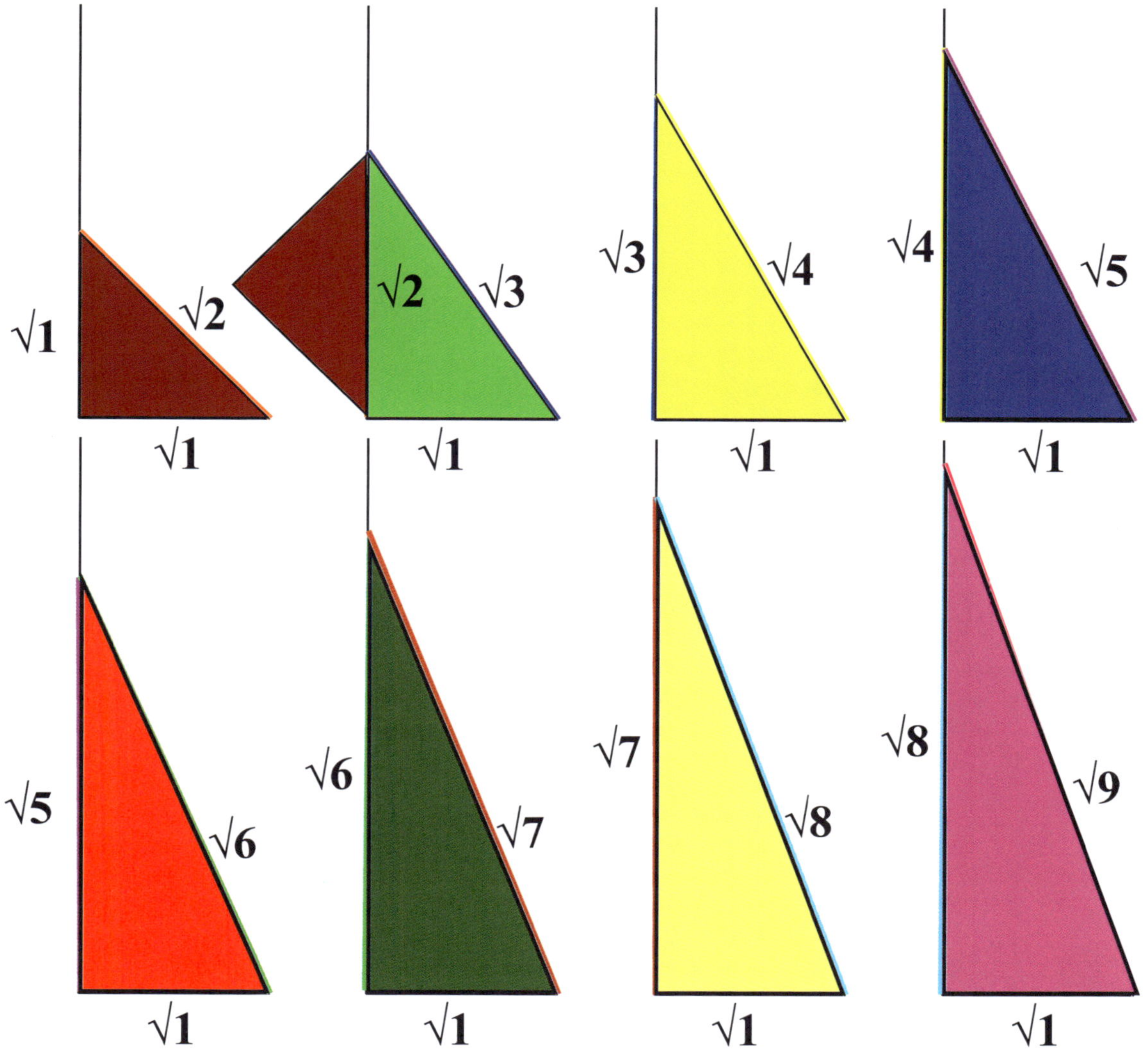

The 1,1,√2 triangle begins a sequence that is credited to Greek Mathematician Theodorus of Cyrene. He lived about 2500 years ago. The 1,1,√2 (√1,√1,√2) triangle is made by taking the diagonal of the Unit Square. Using the hypotenuse of this triangle, √2, as the length of one of the sides next to the 90° angle and the other side with a length of 1, √1, gives a triangle with a hypotenuse of √3. This is the √1,√2,√3 Triangle. Repeat with the successive triangles and place them next to each other and you have the Spiral of Theodorus.

The first triangle is the diagonal of the 1x1 square.

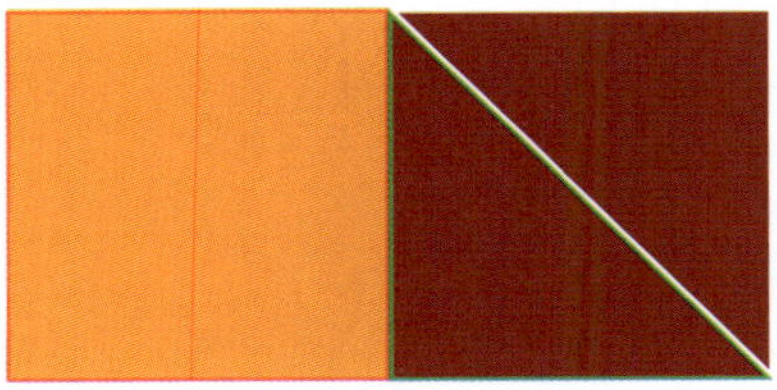

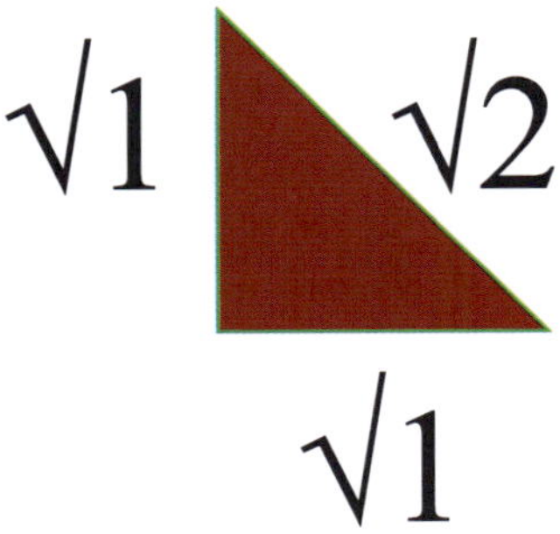

For our next triangle we take the $\sqrt{1}, \sqrt{1}, \sqrt{2}$ triangle and place it along the sides of a one inch wide strip. We then draw a line to mark our desired length, $\sqrt{2}$. This gives us a rectangle, $\sqrt{1} \times \sqrt{2}$. We draw the diagonal and have two $\sqrt{1}, \sqrt{2}, \sqrt{3}$ triangles.

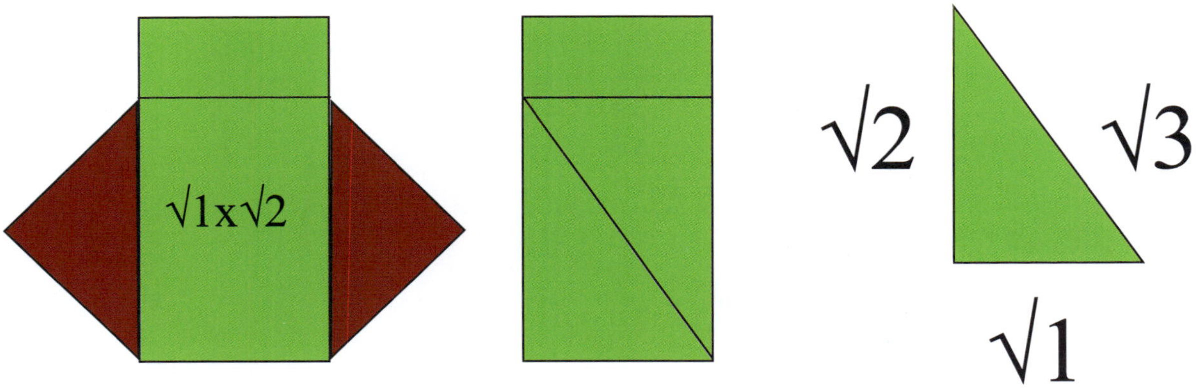

We take this triangle and repeat the process. We lay two triangles along side of a one inch wide strip, draw our rectangle, and then take the diagonal.

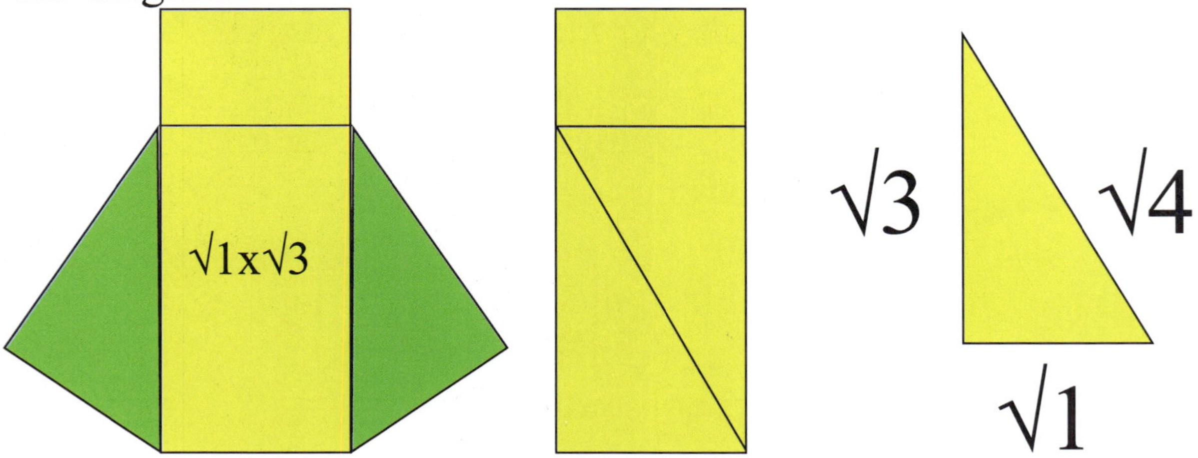

We repeat the process again. We lay two triangles along side of a one inch wide strip, draw our rectangle, and then take the diagonal. This gives us our last triangle and together they form the first four triangles of the Spiral of Theodorus.

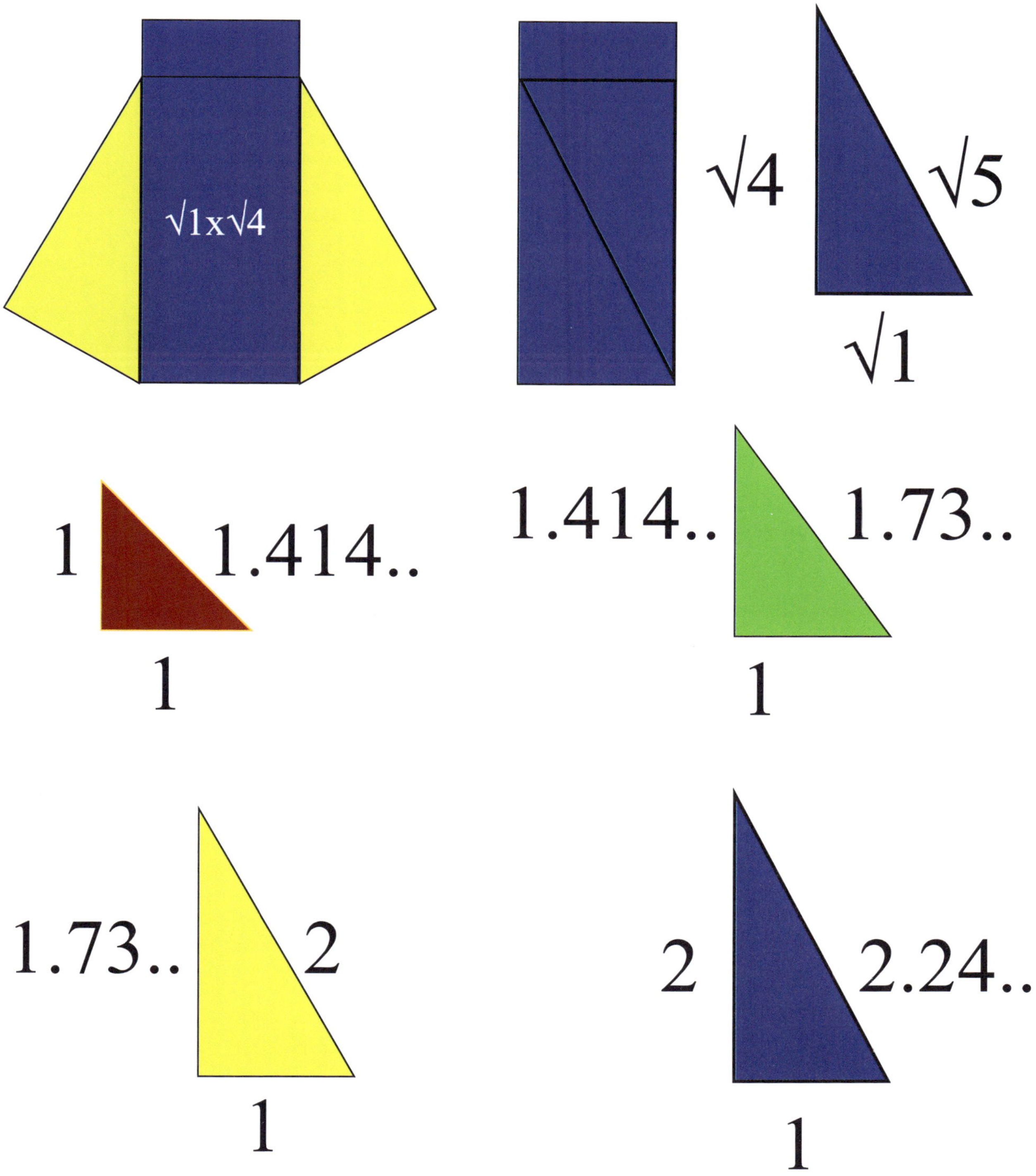

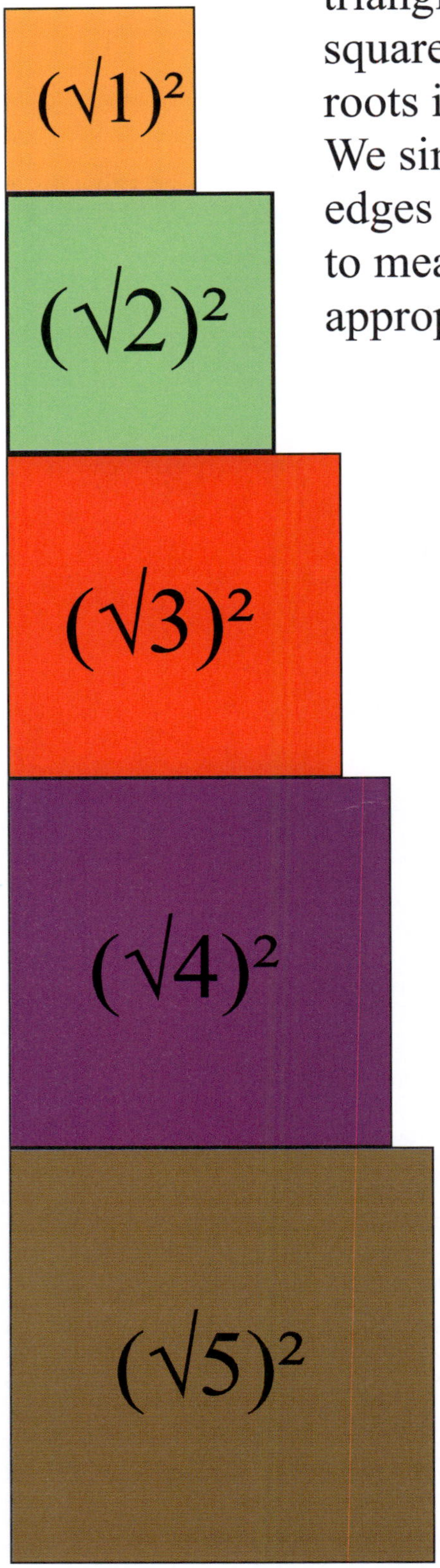

Once we have the triangles, getting the squares of the square roots is fairly easy. We simply use the edges of the triangles to measure the appropriate length.

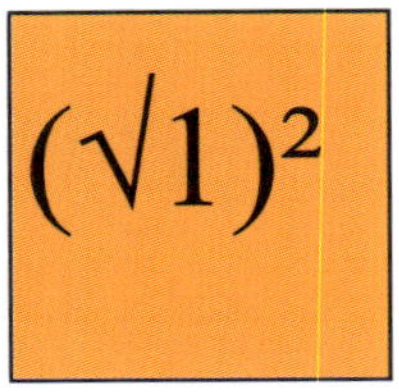

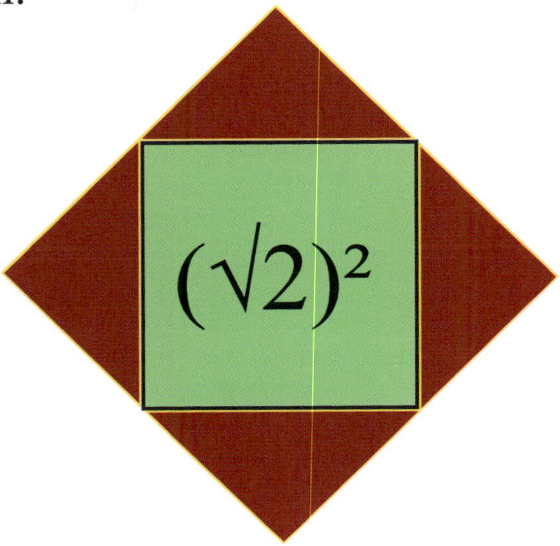

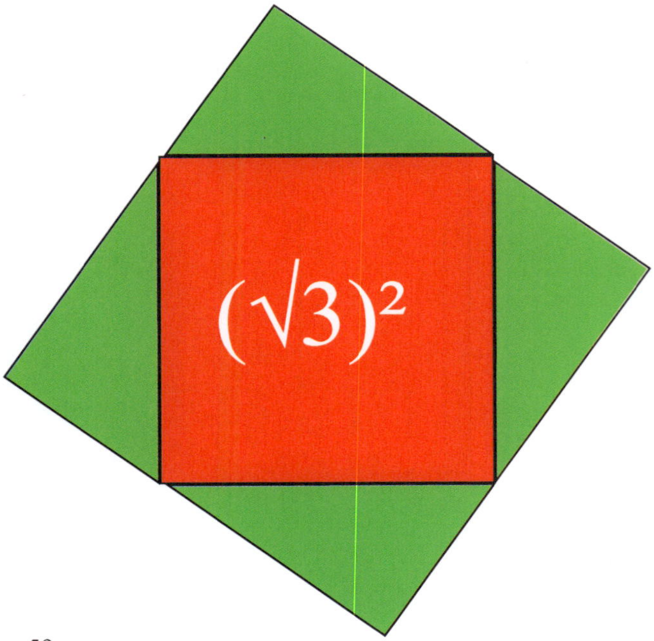

Or we turn the hypotenuse outward
and make the square roots squared.

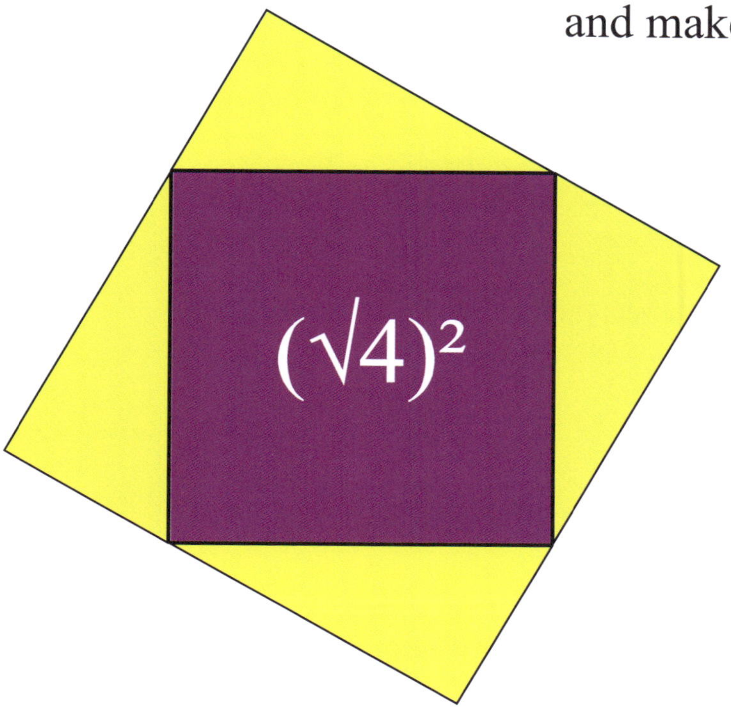

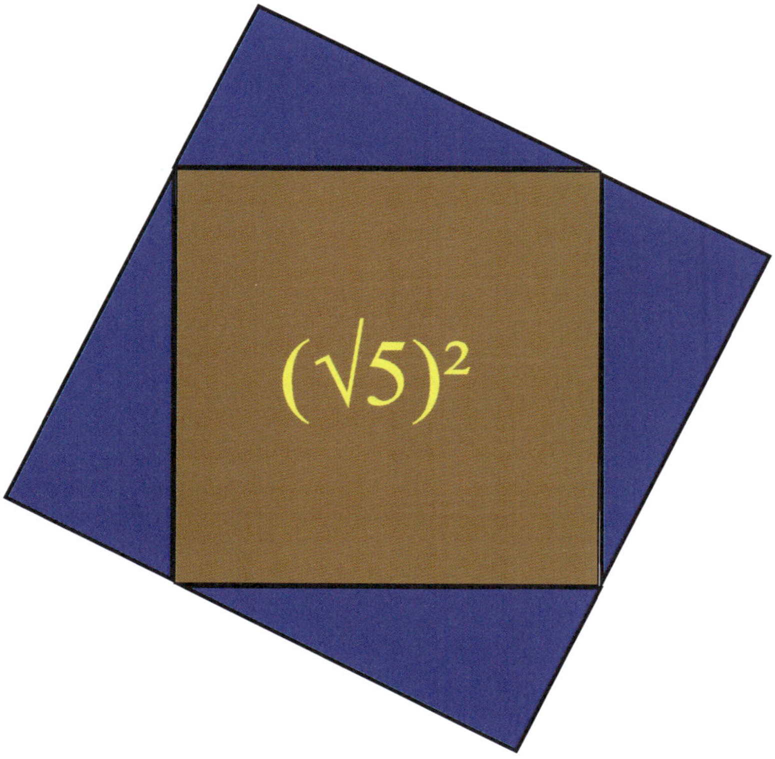

The Square Roots Squared

$\sqrt{1} \times \sqrt{1} = 1$
$\sqrt{2} \times \sqrt{2} = 2$
$\sqrt{3} \times \sqrt{3} = 3$
$\sqrt{4} \times \sqrt{4} = 4$
$\sqrt{5} \times \sqrt{5} = 5$

The first five square roots squared are:

$\sqrt{1} = 1$
$\sqrt{2} \approx 1.41..$
$\sqrt{3} \approx 1.74..$
$\sqrt{4} = 2$
$\sqrt{5} \approx 2.24..$

$$\sqrt{1} \times \sqrt{1} = 1$$
$$\sqrt{1} = 1$$

$\sqrt{1}$

$$\sqrt{2} \times \sqrt{2} = 2$$
$$\sqrt{2} = 1.414$$

$\sqrt{2}$

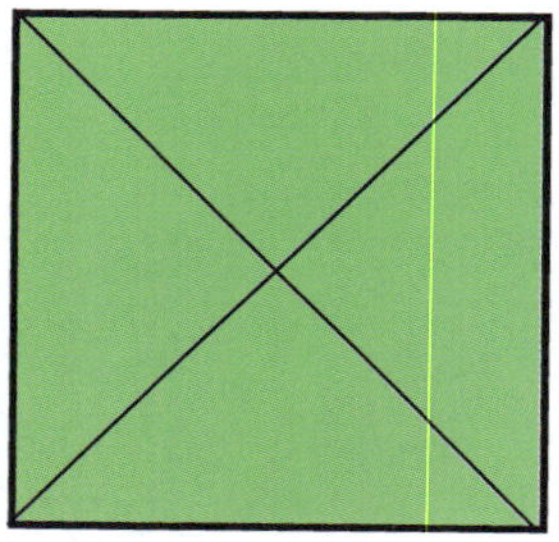

$$\sqrt{3} \times \sqrt{3} = 3$$
$$\sqrt{3} = 1.732..$$

$\sqrt{3}$

$\sqrt{4}$

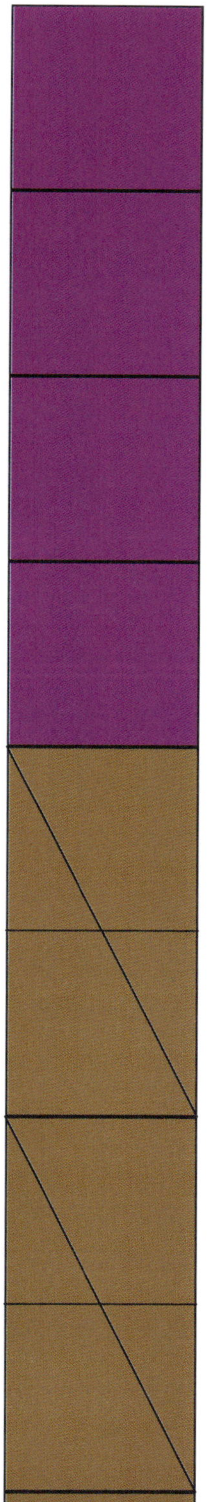

$$\sqrt{4} \times \sqrt{4} = 4$$
$$\sqrt{4} = 2$$

A square root is really a question. If you know the number of 1x1 squares in your larger square, what are the lengths of the sides of the larger square that will give you that number of 1x1 squares?

$\sqrt{5}$

$$\sqrt{5} \times \sqrt{5} = 5$$
$$\sqrt{5} = 2.236..$$

$\sqrt{100}$ is 10 because you would need a square with sides of 10 to give you 100 1x1 squares. This is one of the <u>Whole Number Squares</u>.

$(\sqrt{1})^2 = 1$
$(\sqrt{2})^2 = 1 + 1 = 2$
$(\sqrt{3})^2 = 1 + 1 + 1 = 3$
$(\sqrt{4})^2 = 1 + 1 + 1 + 1 = 4$
$(\sqrt{5})^2 = 1 + 1 + 1 + 1 + 1 = 5$

See how the square roots squared grow.
$(\sqrt{2})^2$ is 1 more than $(\sqrt{1})^2$
$(\sqrt{3})^2$ is 1 more than $(\sqrt{2})^2$
$(\sqrt{4})^2$ is 1 more than $(\sqrt{3})^2$
$(\sqrt{5})^2$ is 1 more than $(\sqrt{4})^2$

$(\sqrt{1})^2 = 1 \text{ x } 1 = 1$

$(\sqrt{2})^2 \approx 1.41.. \text{ x } 1.41.. = 2$
$\approx$ equals approximately

$1 + 0.41.. \approx 1.41..$

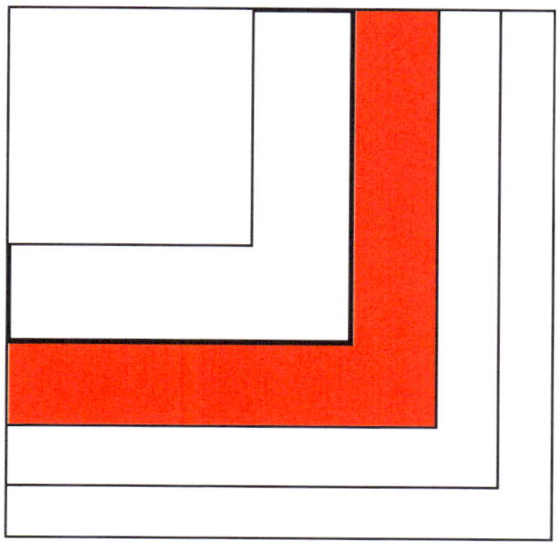

$1.41.. + 0.32.. \approx 1.73..$

$(\sqrt{3})^2 \approx 1.73.. \times 1.73.. = 3$

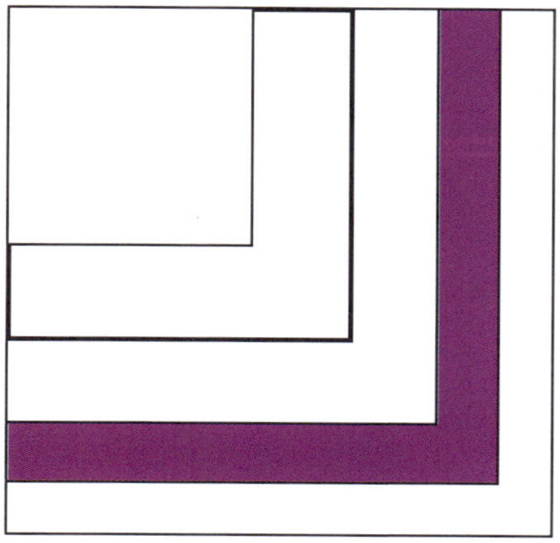

$1.73.. + 0.27.. = 2.00$

$(\sqrt{4})^2 = 2 \times 2 = 4$

$2.00 + 0.24.. \approx 2.24..$

$(\sqrt{5})^2 = 2.24.. \times 2.24.. \approx 5$

The Pythagorean Theorem

The 1, √2, √3 Triangle

$a^2 + b^2 = c^2$ $\qquad$ $1^2 + (\sqrt{2})^2 = (\sqrt{3})^2$

$1 + 2 = 3$

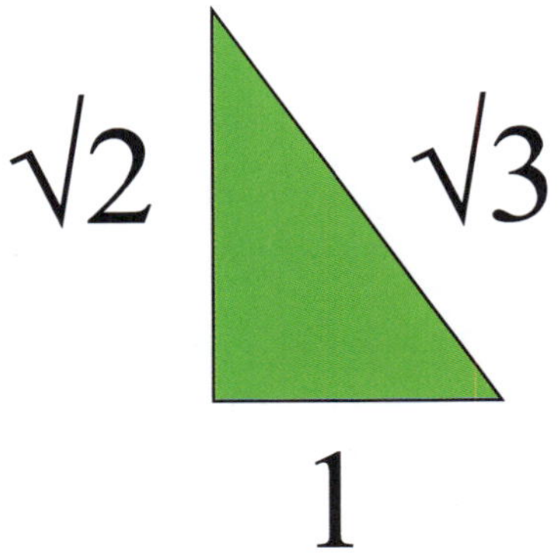

This triangle is the second one in the spiral of Theodorus. It has no convenient demonstration so we must depend on our proofs to establish its validity.

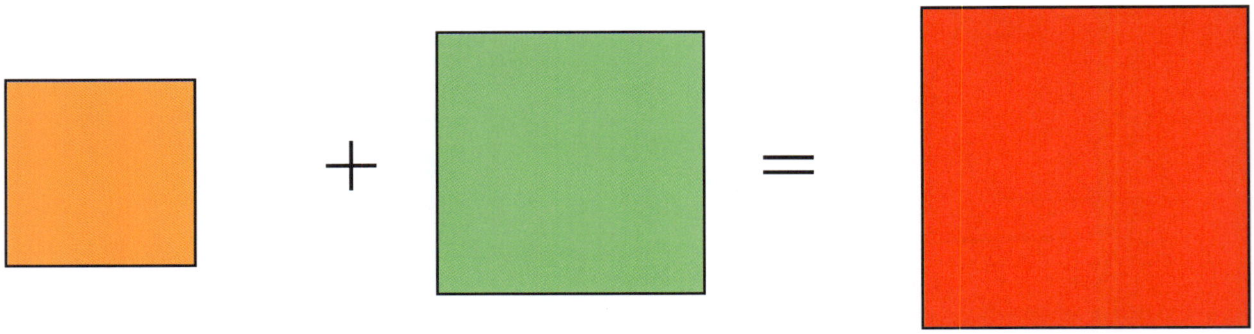

We create two equal squares. One with the hypotenuse squared and four triangles and the other with the other two sides squared and four triangles. Take the triangles away and what is left is equal.

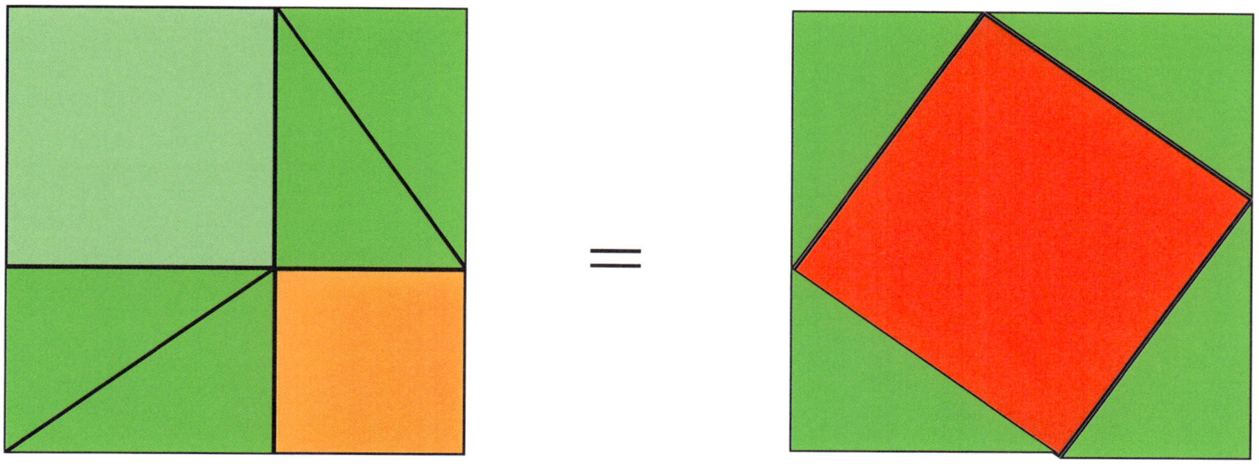

Our other proof is using the triangles to make the square of the hypotenuse and then moving two of the triangles to create a shape equal to the other two sides squared.

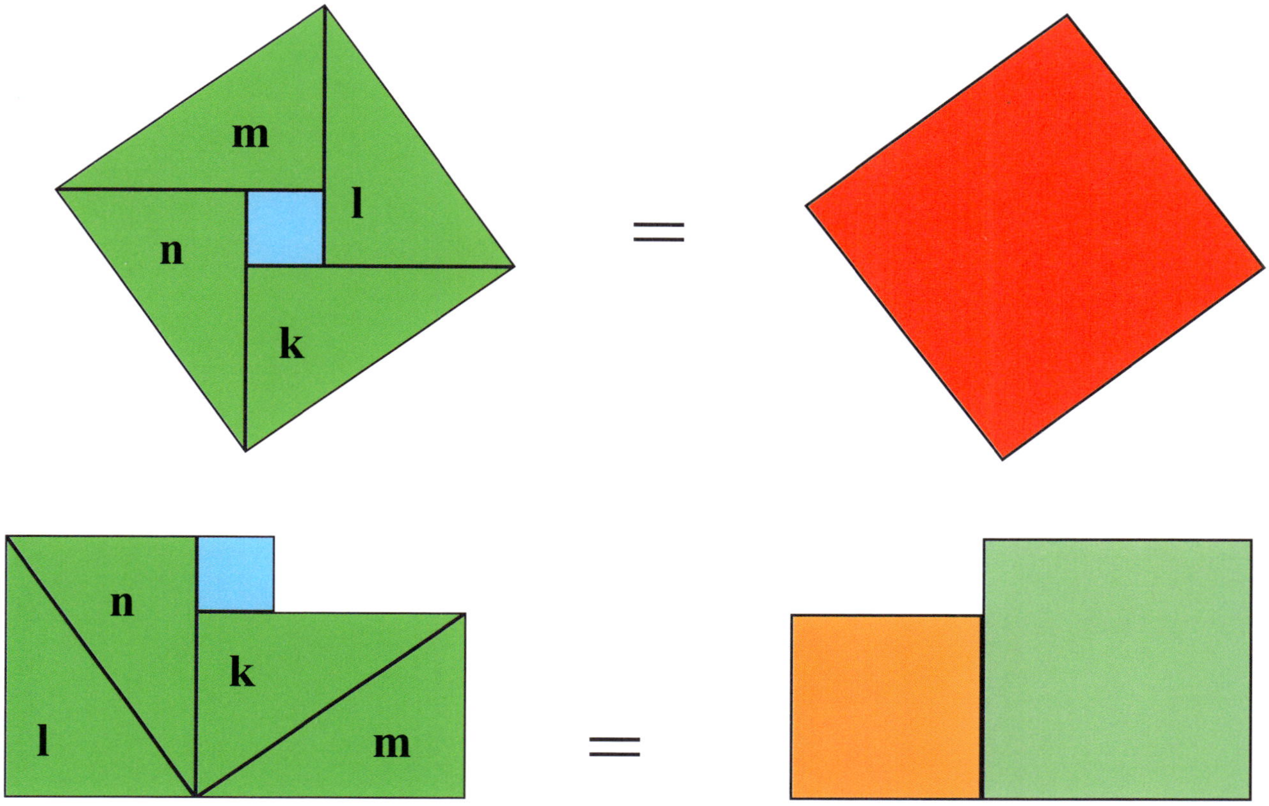

The Pythagorean Theorem

The 1, √3, 2 Triangle   The 30°, 60°, 90° Triangle

$a^2 + b^2 = c^2$ $1^2 + (\sqrt{3})^2 = 2^2$ $1 + 3 = 4$

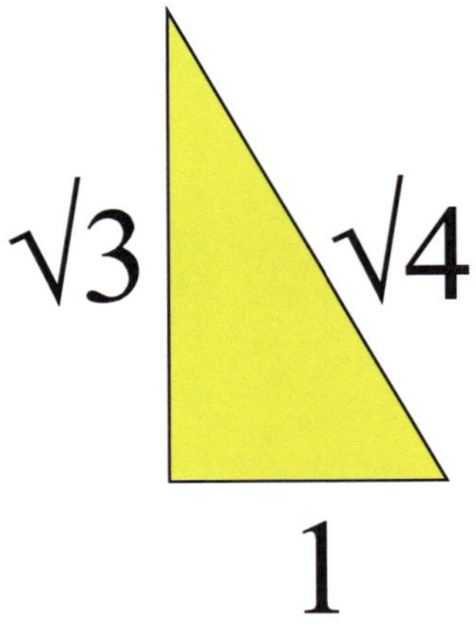

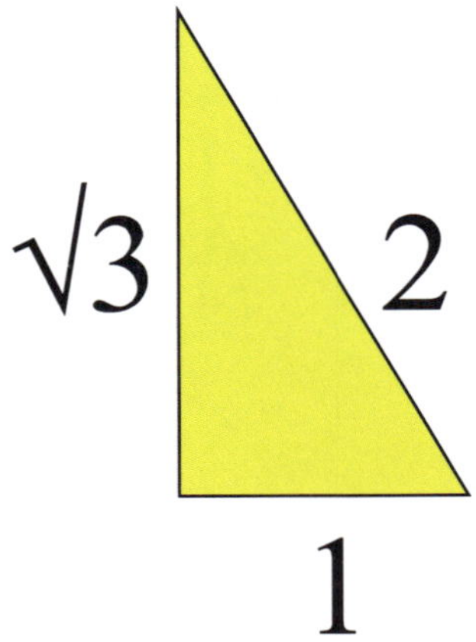

This is the third triangle in the spiral of Theodorus. It has no convenient demonstration so we must depend on our proofs to establish its validity.

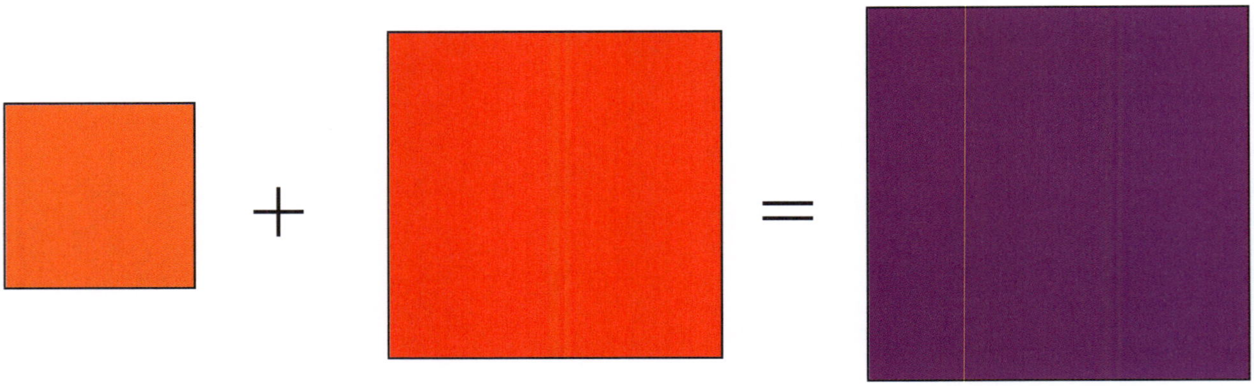

Create two equal squares, one with the hypotenuse squared and four triangles and the other with the other two sides squared and four triangles. Take the triangles away and what's left is equal.

=

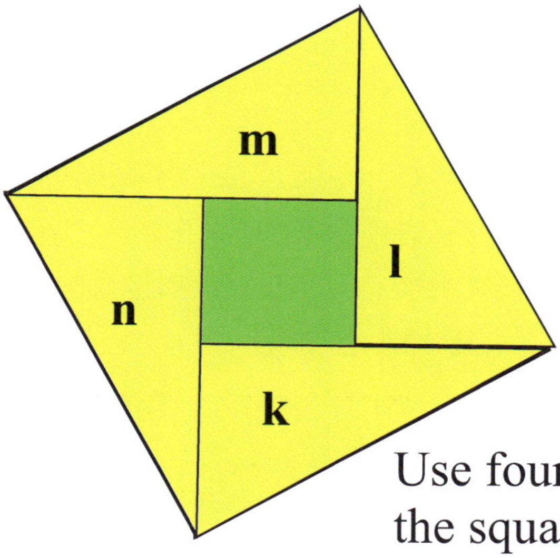

=

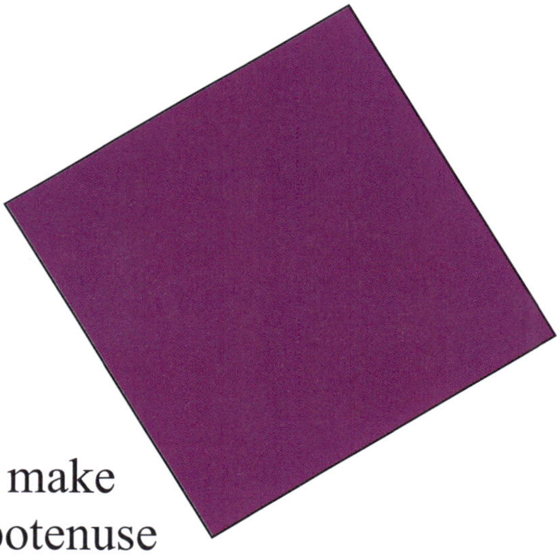

Use four triangles to make the square of the hypotenuse and then move two of the triangles to create a shape equal to the other two sides squared.

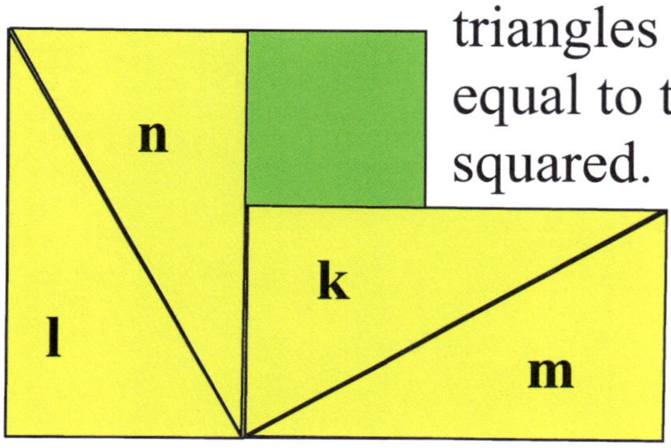

=

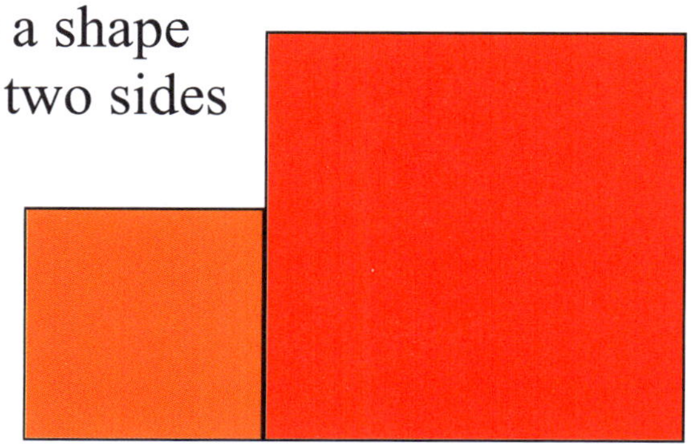

The Pythagorean Theorem

1, 2, √5 Triangle   27°, 63°, 90°

$a^2 + b^2 = c^2$ $1^2 + 2^2 = (\sqrt{5})^2$ $1 + 4 = 5$

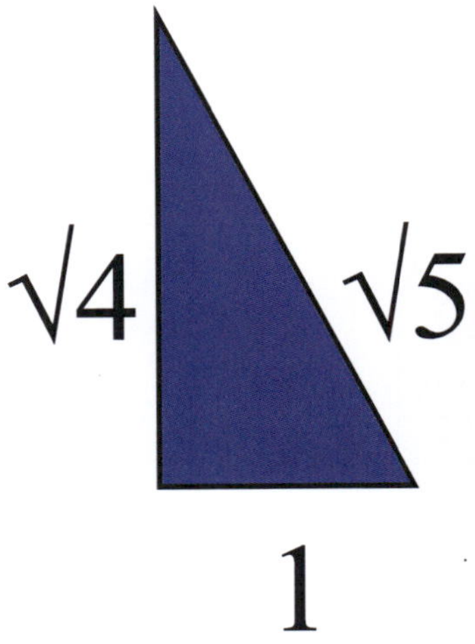

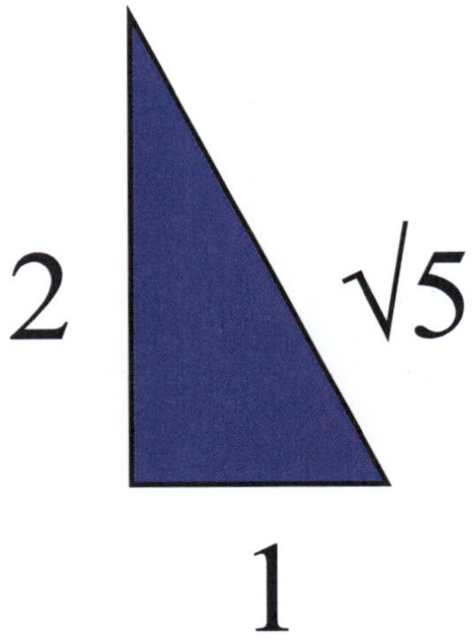

This triangle is quite famous. It even has a name, the Dom. It is called the Dom because it is the shape you get by cutting a Domino, (a 1 x 2 rectangle) along its diagonal. It is used in the Penrose Tiles. (See Web)

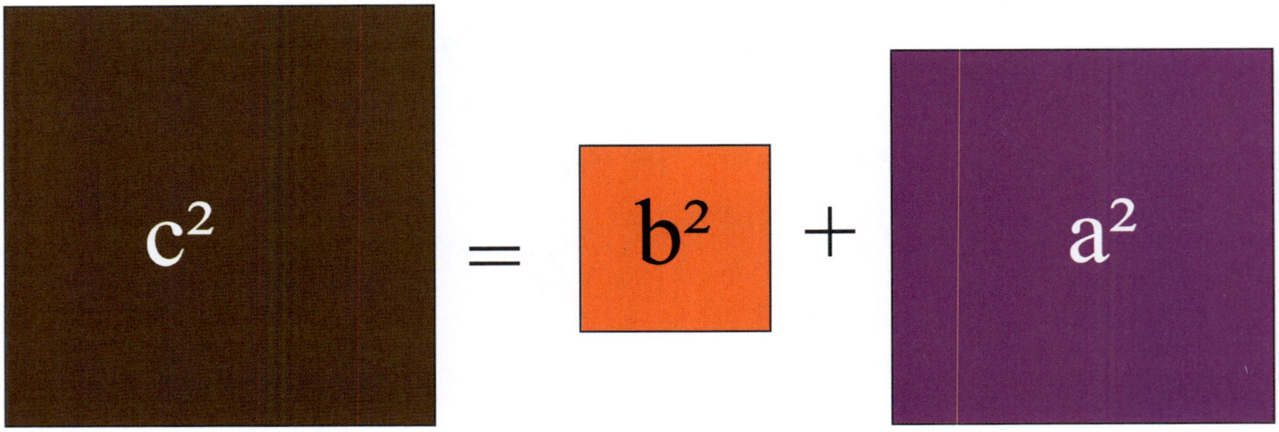

It also has a very elegant special demonstration. We can create the square of the hypotenuse with four triangles and the square of the smallest side. The four triangles can be arranged into the square of the other side. This demonstrates that the square of the hypotenuse is equal to the sum of the squares of the other two sides.

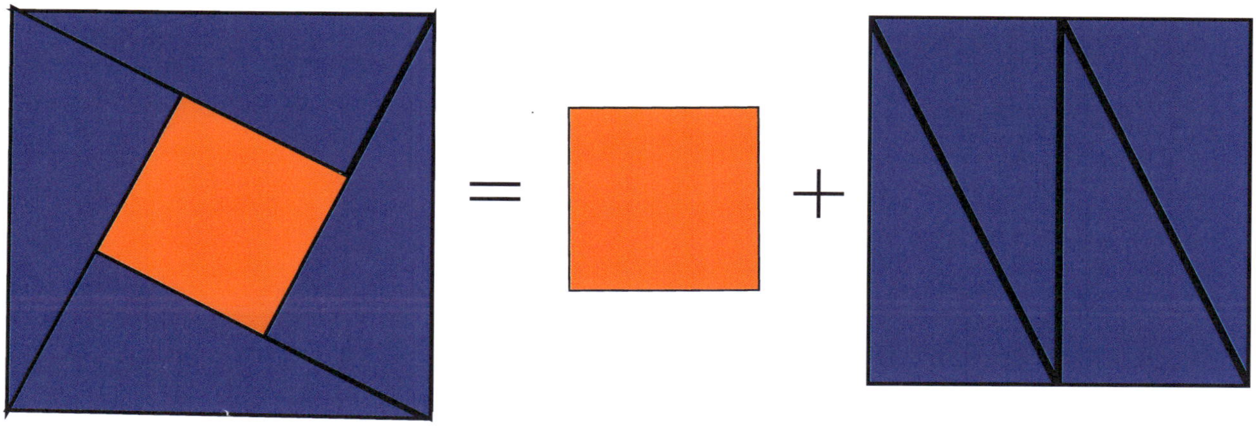

And now to the proofs. Create two equal squares, one with the hypotenuse squared and four triangles arrayed around it and the other square comprised of the other two sides squared and four triangles arranged into two rectangles. Take the triangles away and what's left is equal.

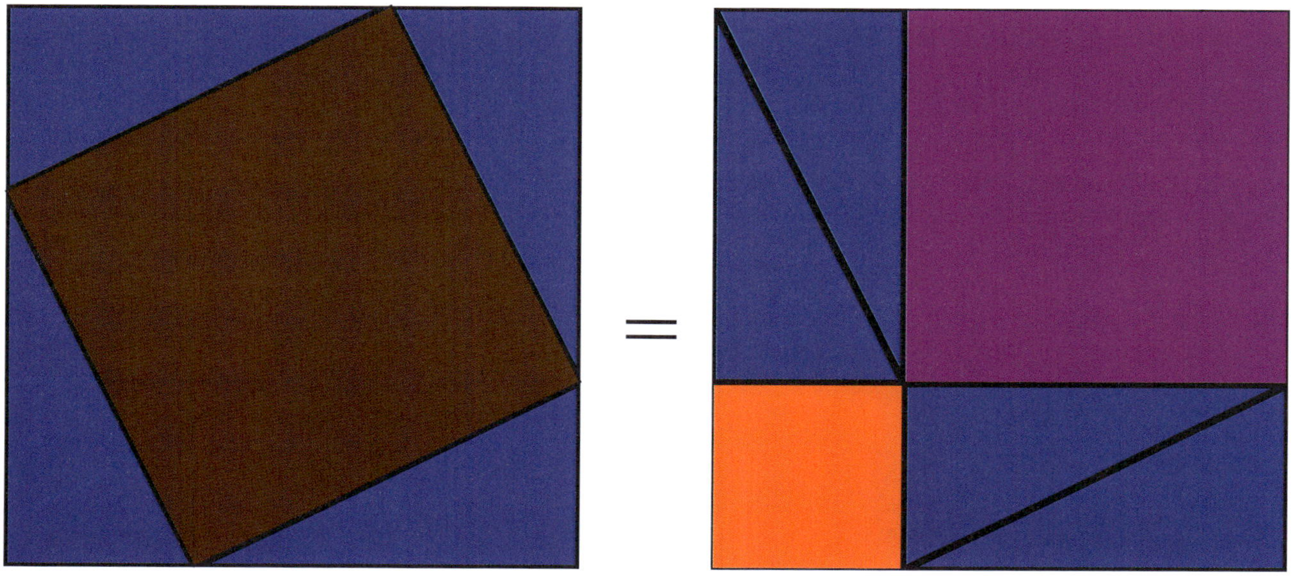

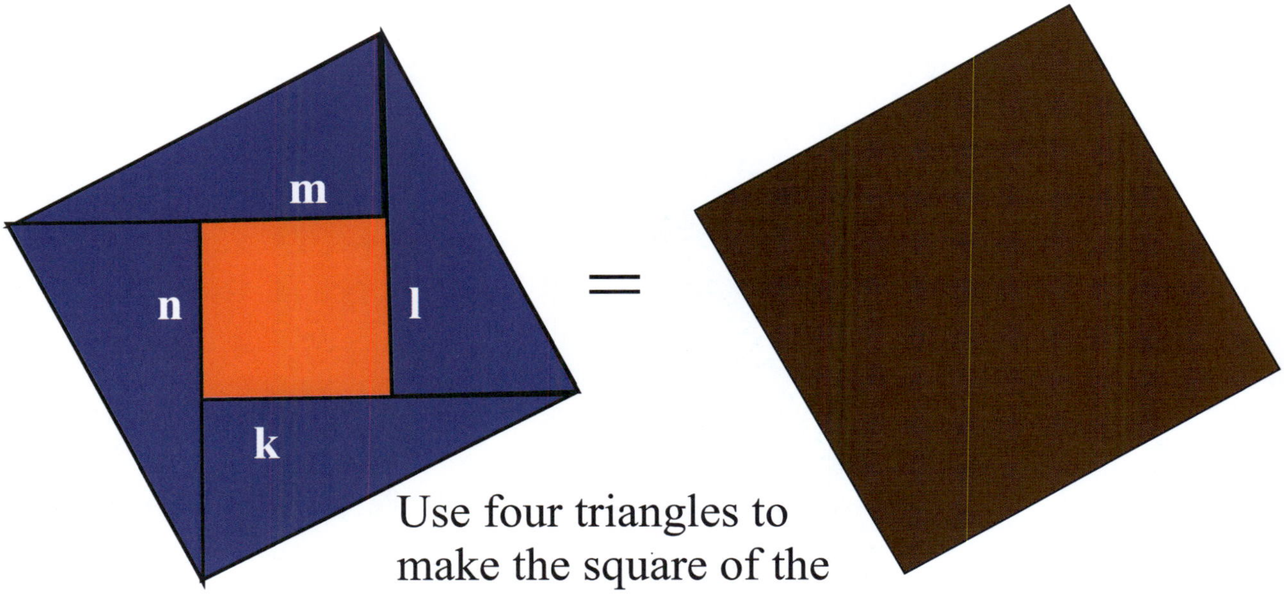

Use four triangles to make the square of the hypotenuse and then move two of the triangles to create a shape equal to the other two sides squared.

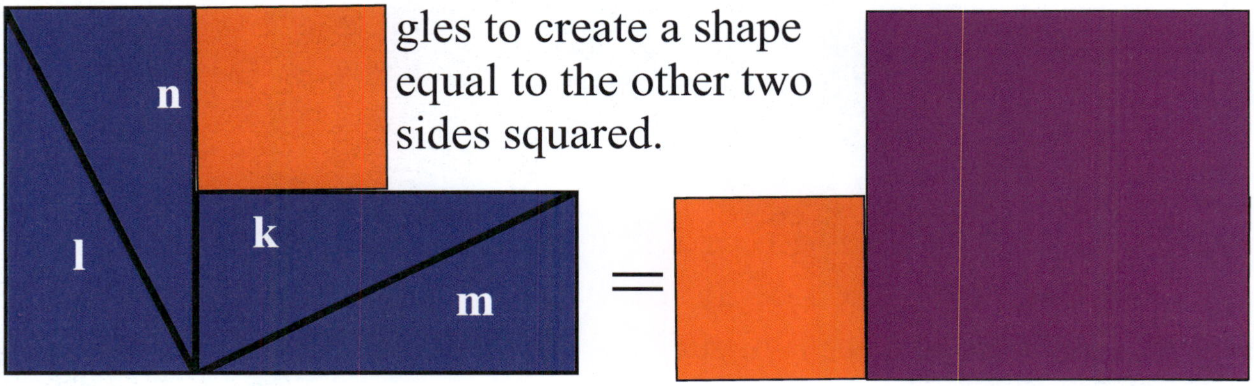

We like this shape so much that we made it the logo for our company <u>Math for the Ages</u>. You can just keep putting triangles around it and all of the rectangles that are constructed are Golden Rectangles, including the one in the middle.

The Draftsman's Pair

45°, 45°, 90° and 30°, 60°, 90°

You probably recognize these two triangles. You might already own them. They are used in the area of mechanical drawing also known as drafting. This is a process used by engineers, architects and artists who are creating drawings of their plans and projects. They are often sold as a set along with the T-square and drawing board. They can be purchased cheaply at most stores that sell school and office supplies. If you are interested in the process of design, these tools are a good place to start. They will help you to be clean and precise in your drawing efforts.

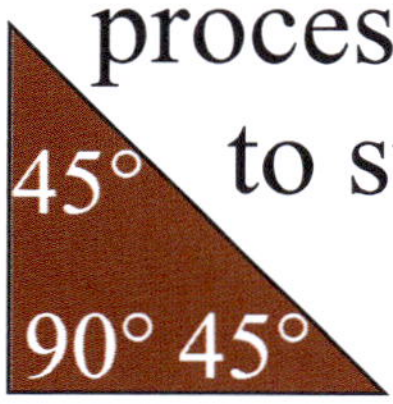

The Interiors
of the Platonic Solids

The interiors of the Platonic Solids are comprised of triangles that connect the focus with two vertices of the various solids. These triangles when put together in groups of 3, 4 and 5 form pyramids that combine to create the Platonic Solids.

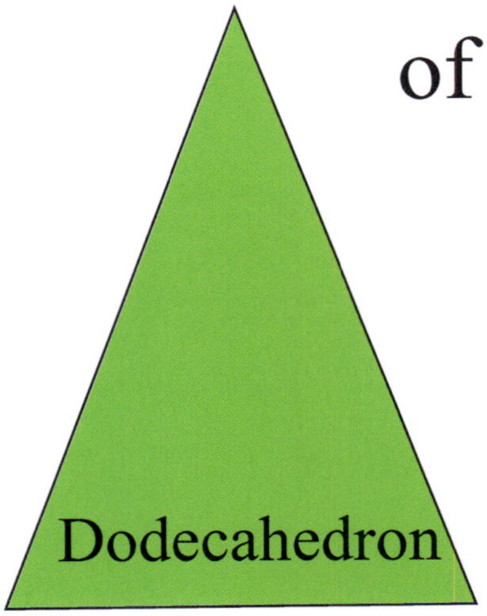

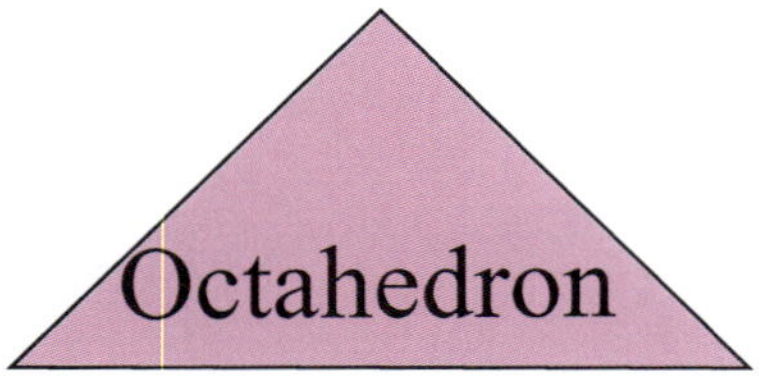

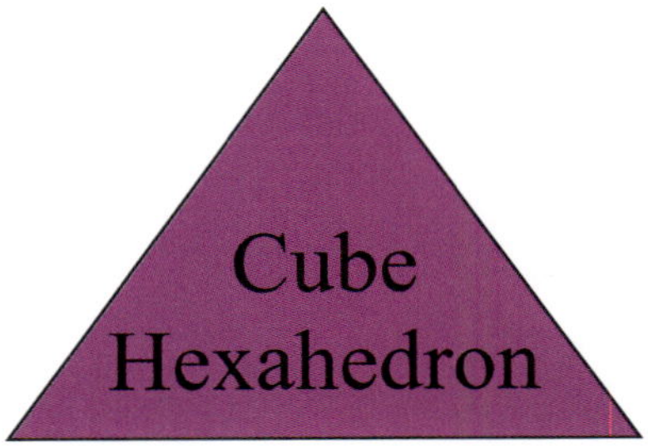

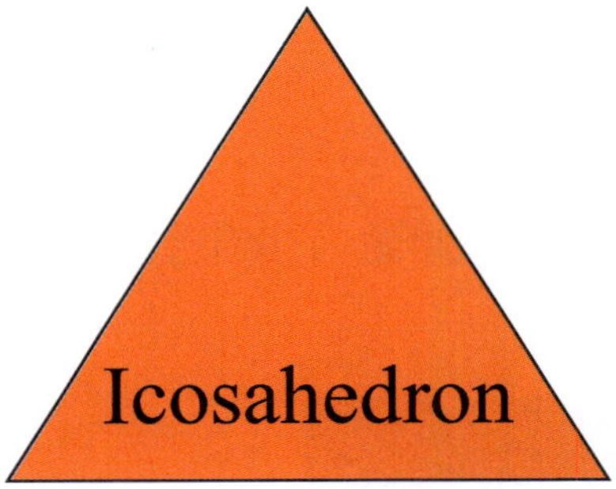

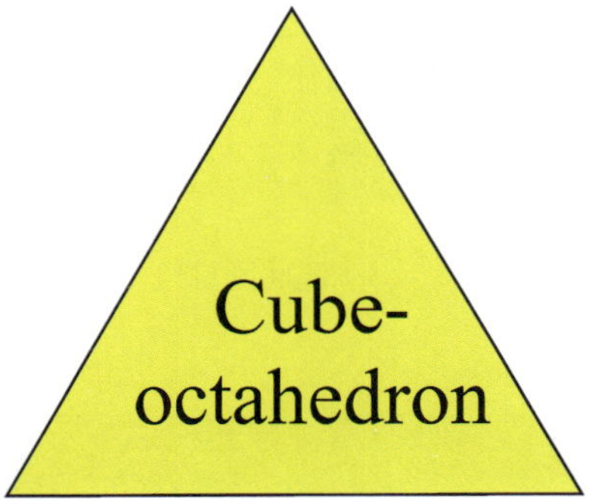

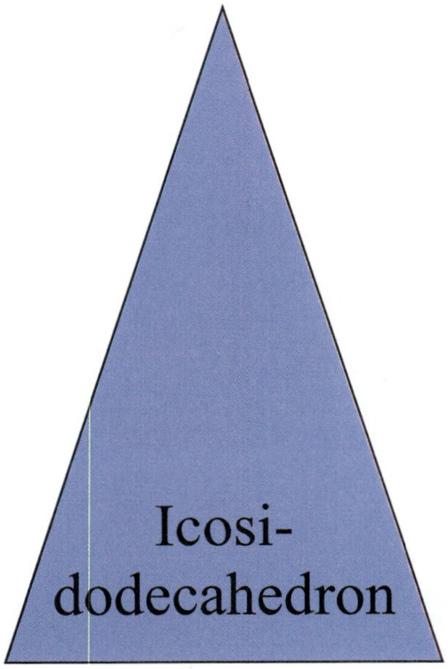

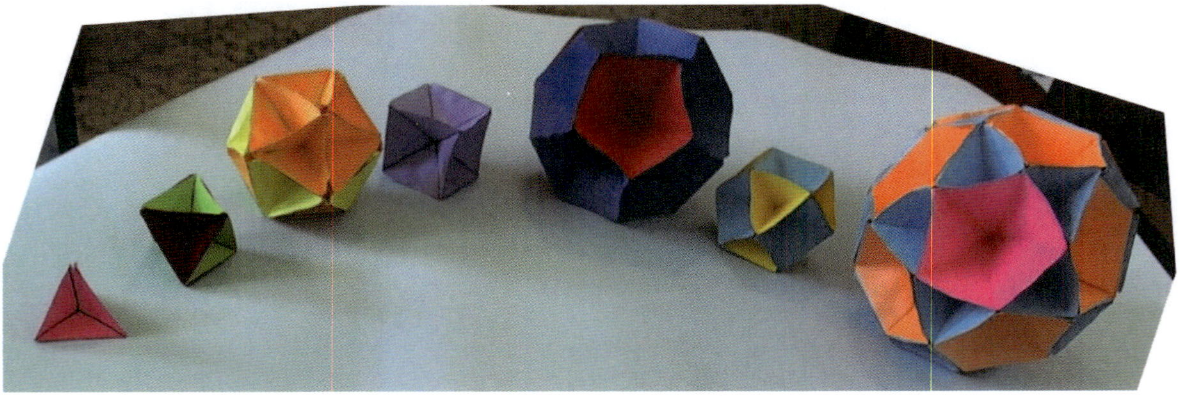

Tetrahedron Interior

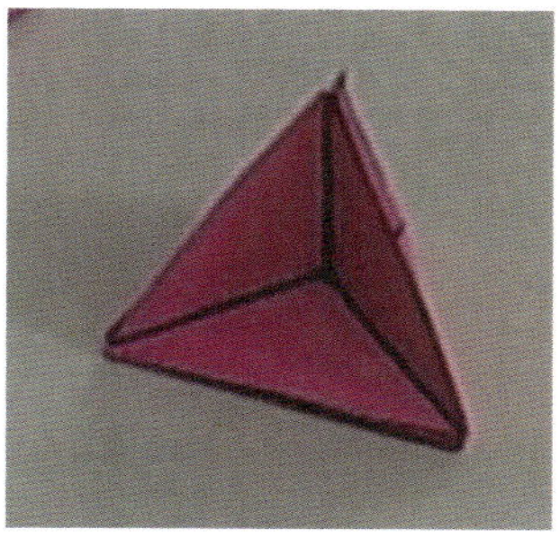

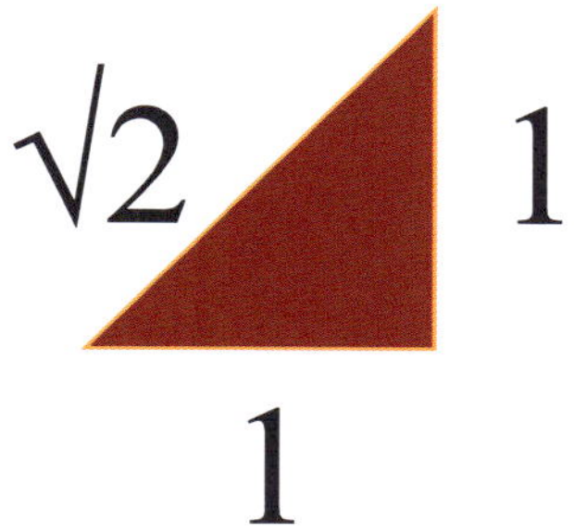

One of the major discoveries since the first edition of this book is the derivation of the measurements of the interiors of the Platonic Solids. We found that each of these triangles can be constructed using the first four triangles of the Spiral of Theodorus. Our first example is the Tetrahedron.

We arrange some 1,1,√2 triangles into the rectangle below.

We then trace the rectangle and draw in the diagonals.

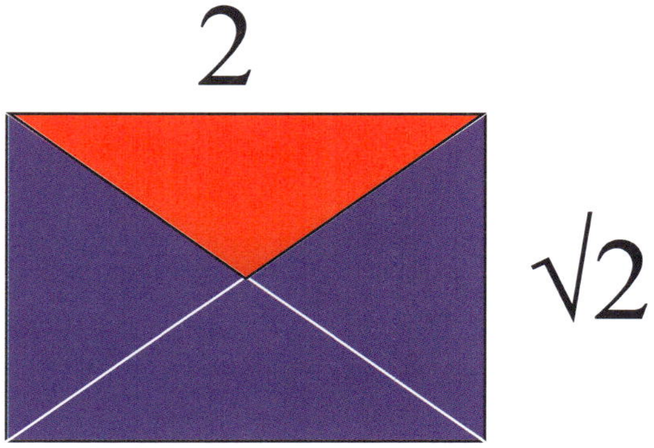

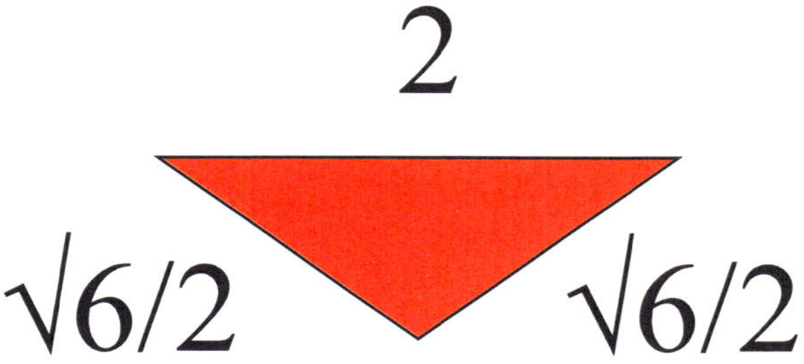

This gives us this triangle.

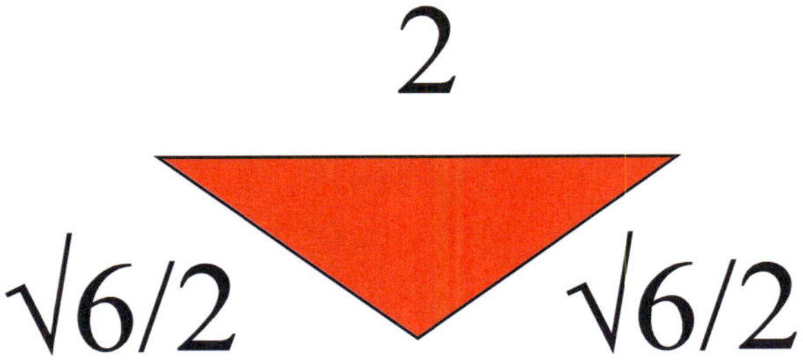

Each of the internal pyramids is made of three of these triangles. Cut them out of paper and fold each line and attach them to each other with the small flap.

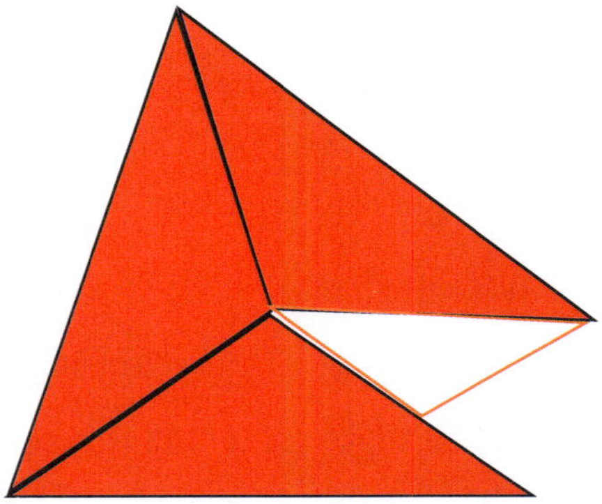

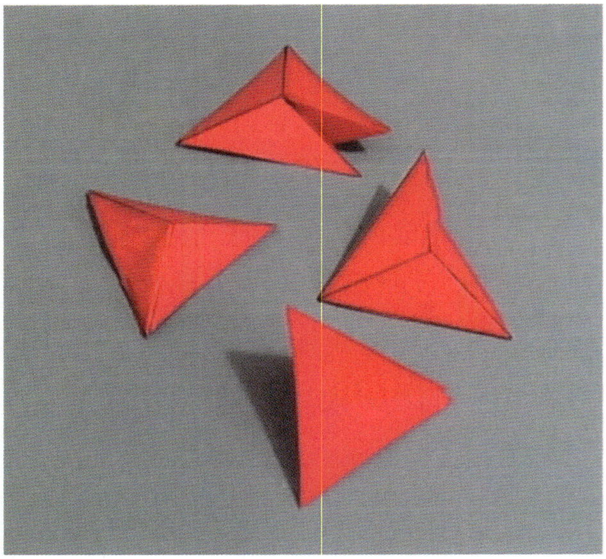

Make four of these little pyramids and glue them together and you have a Tetrahedron.

Octahedron Interior

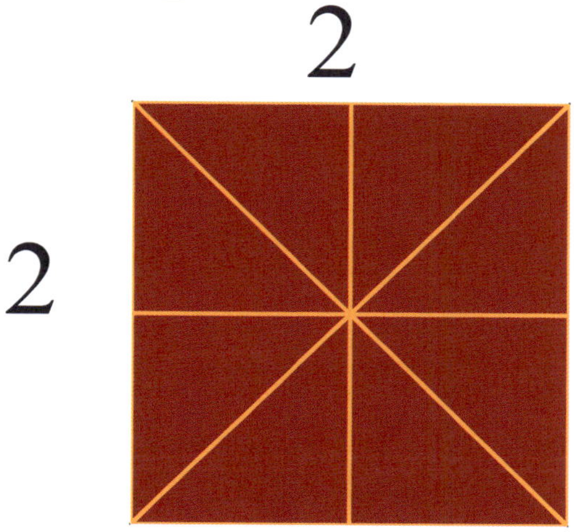

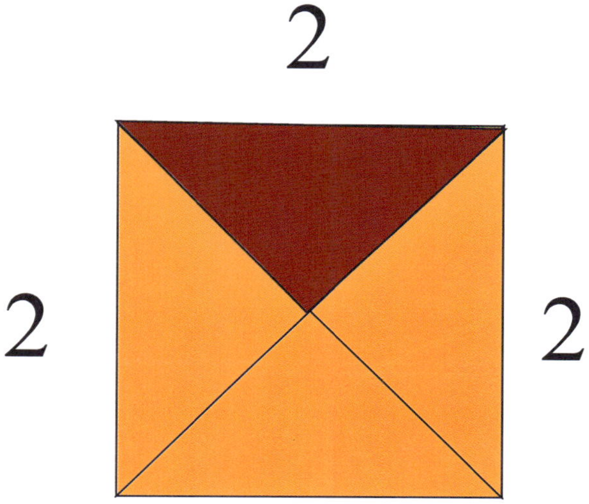

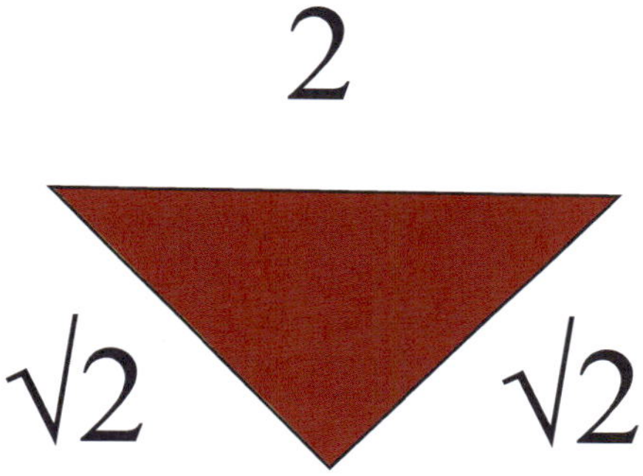

The Octahedron can be made by creating a square and folding it along both of its diagonals. Then just as we did with the hexagon earlier, you cut one of the diagonals to the middle. You then fold one under and glue the flap.

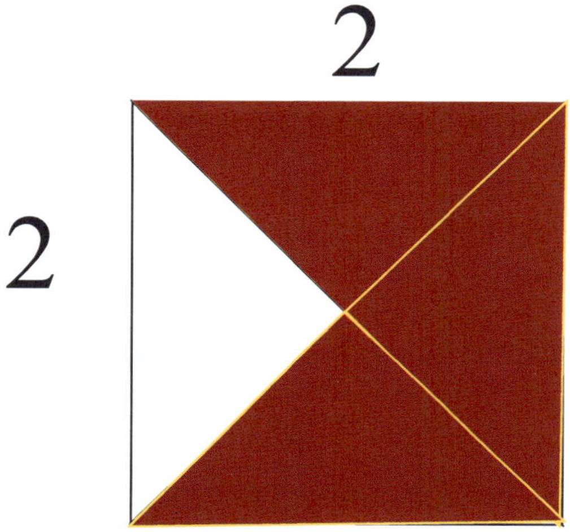

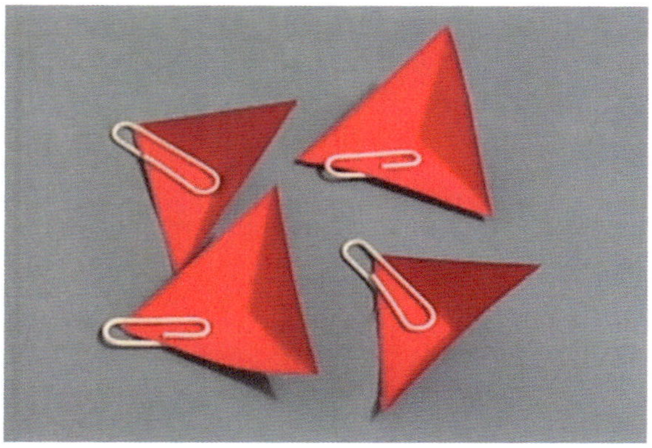

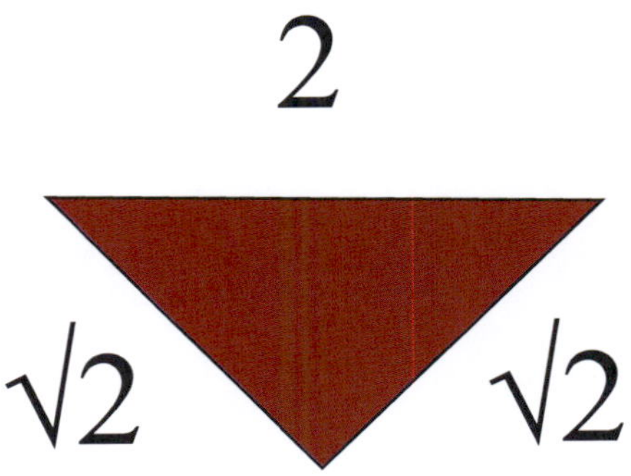

You take eight of these and put them together and you have an octahedron.

This shape gives us a complete intersection of the regular square and the regular triangle.

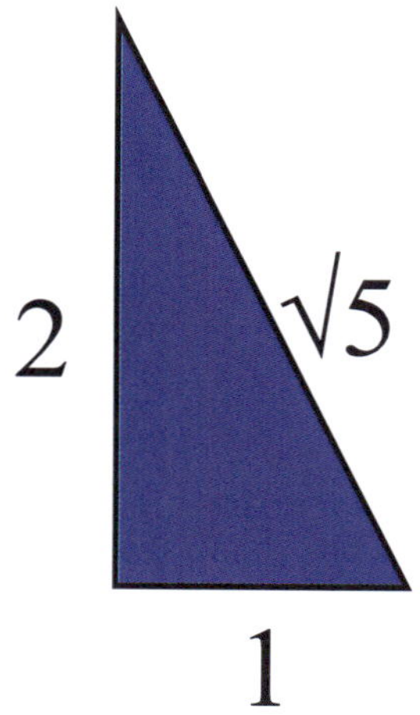

Dodecahedron Interior

The Dodecahedron brings us back to our golden rectangle. We form the Golden Rectangle and place a square underneath it. This forms a new rectangle.

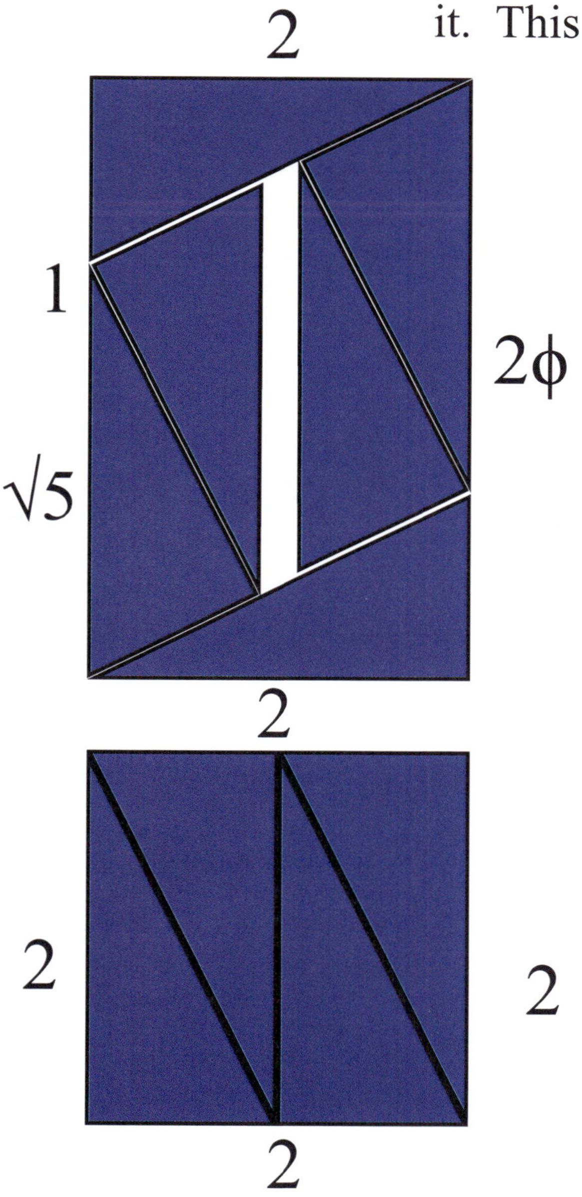

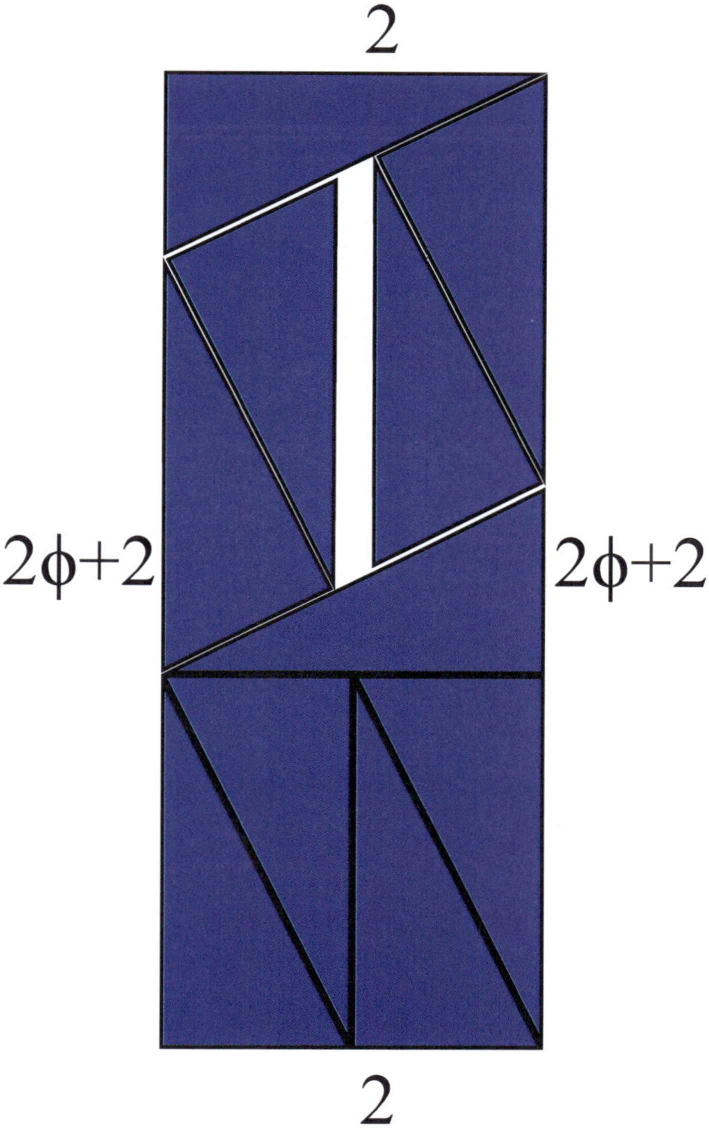

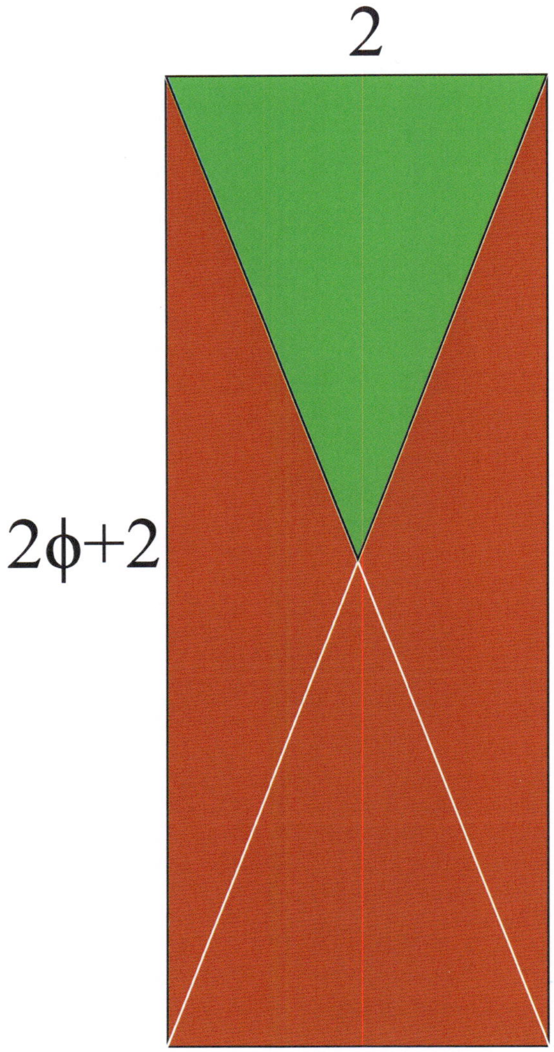

Take the diagonals of this new triangle and that forms the interior triangle of the Dodecahedron

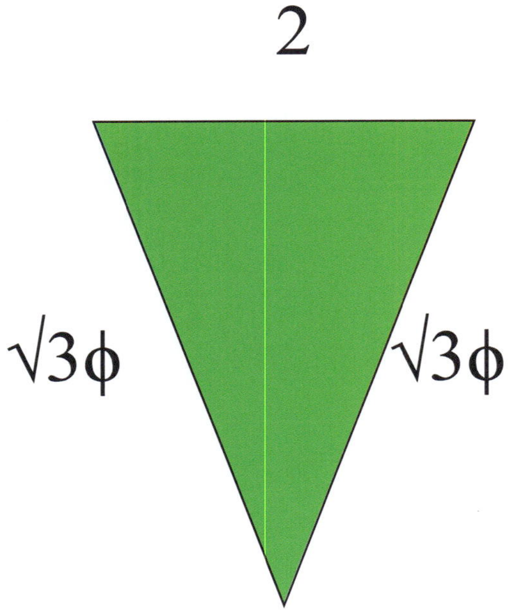

Place six of these triangles together and fold each line. Glue one under and you have the internal pyramid of the Dodecahedron.

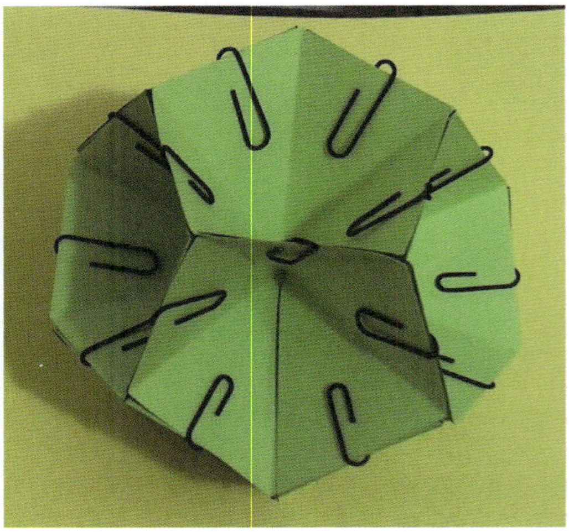

Construct twelve of these pyramids and put them together and you have your Dodecahedron.

The Greeks revered this shape as a representation of the Cosmos.

Hexahedron (Cube) Interior

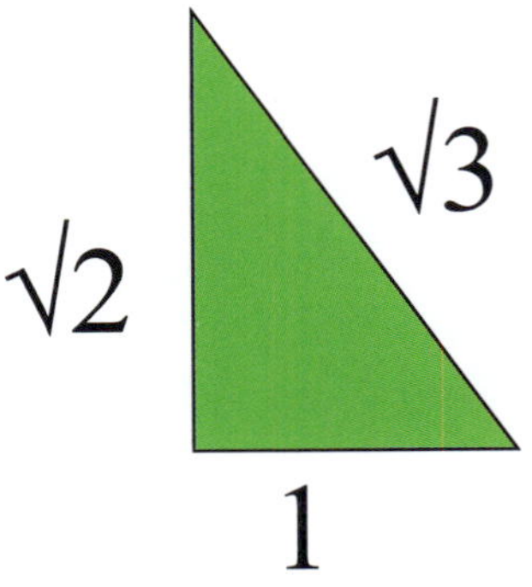

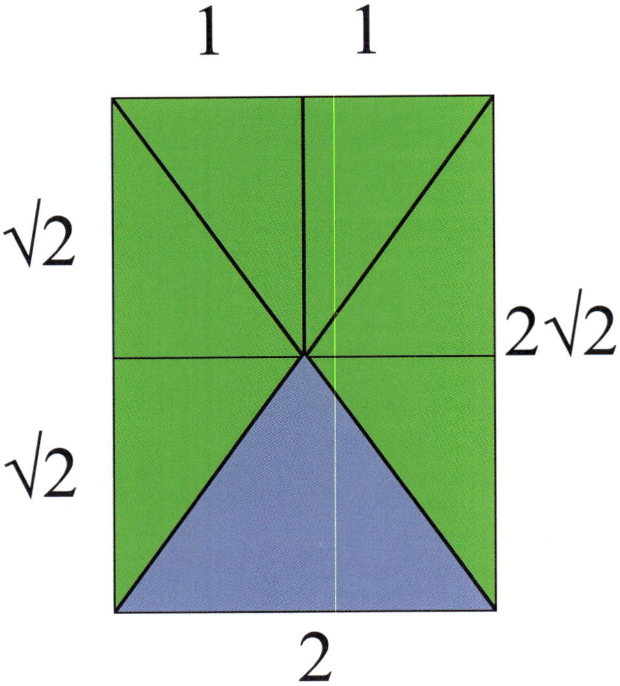

The Cube uses the second triangle of the Spiral of Theodorus to define its interior. Its faces are all squares so we need to array five of our newly created triangles around a center, four for the faces and one as a flap.

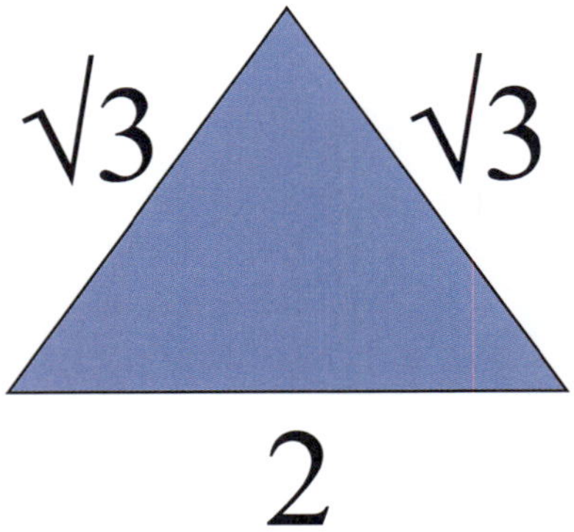

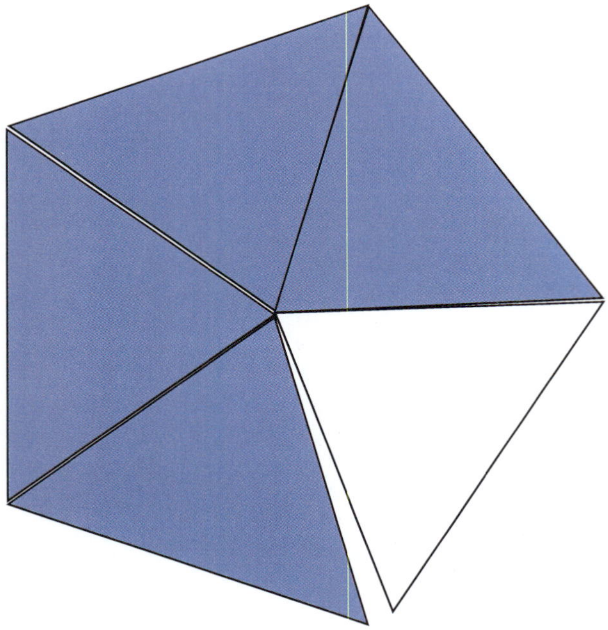

We fold each line and tuck the flap under to make the four sided pyramid. We put six of these together to form the Hexadron (Cube).

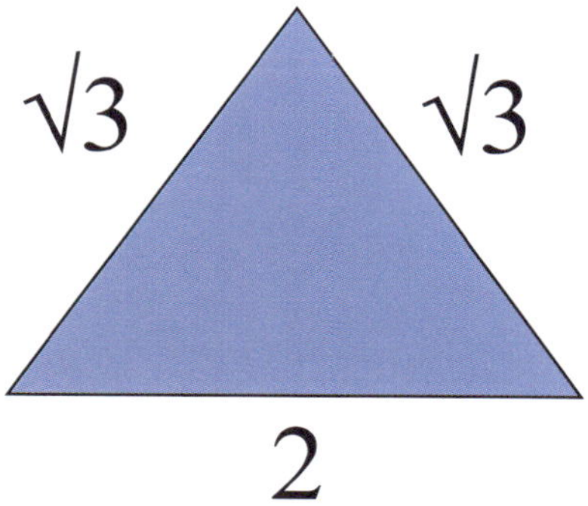

Icosahedron Interior

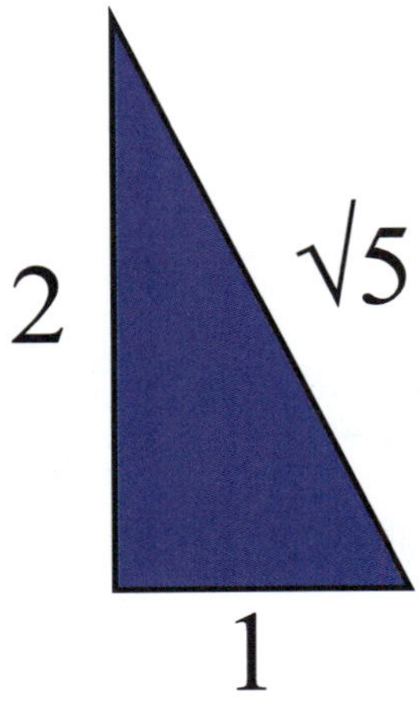

We use the fourth triangle in the Spiral of Theodorus to create the Interior of the Icosahedron.

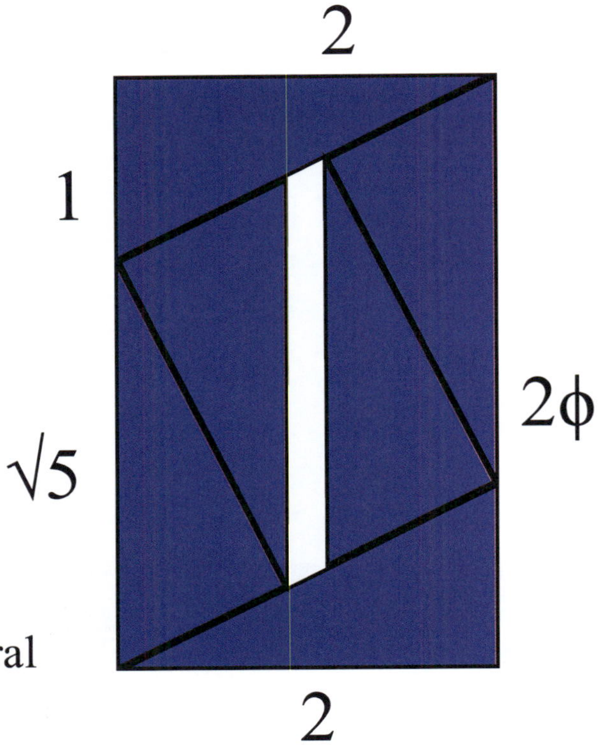

We first construct a Golden Rectangle and then take its diagonal. Using the Pythagorean Theorem and Algebra we calculate its length.

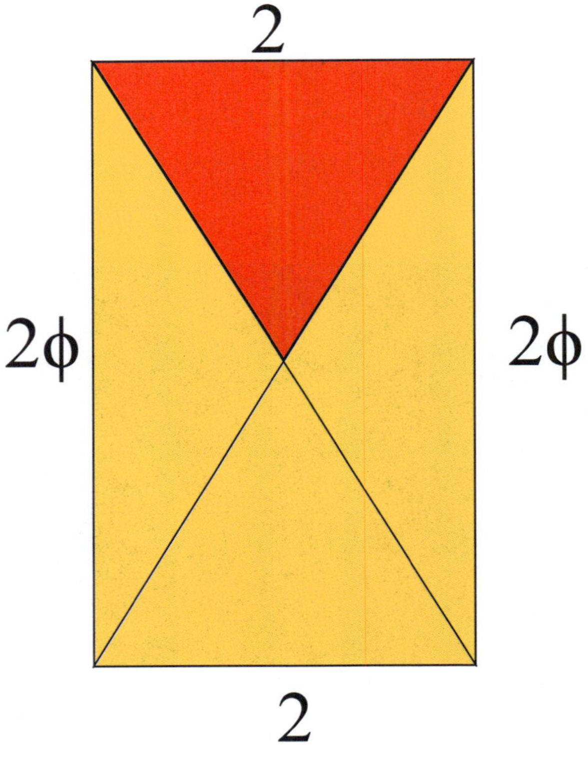

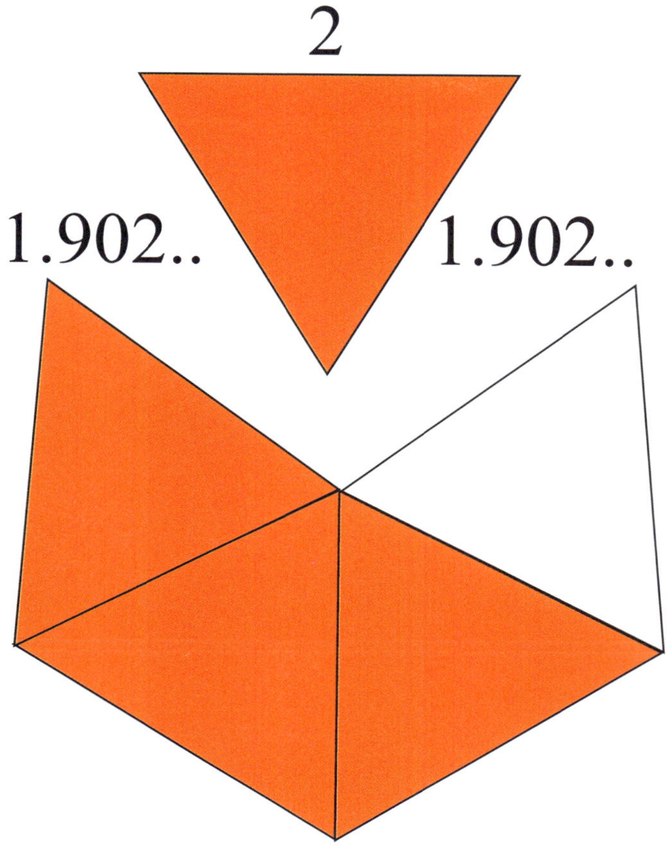

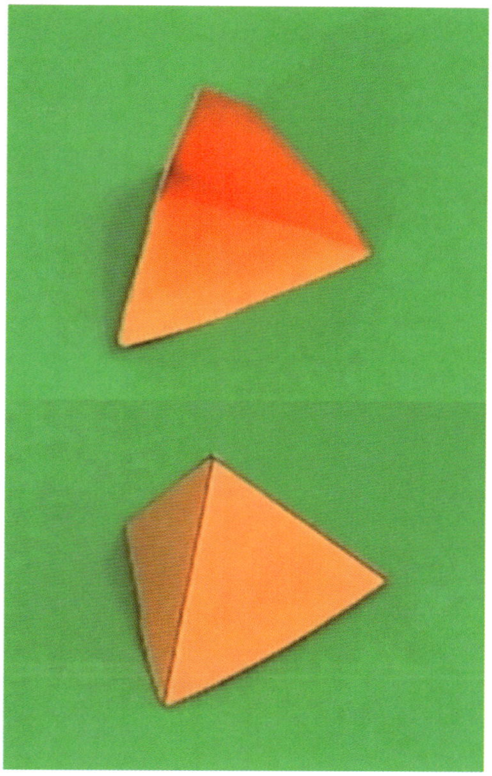

The faces of the Icosahedron are triangles so we construct three sided pyramids. We put together four of our created triangles around a center, leaving one for a flap. We put twenty of these pyramids together and we have the Icosahedron.

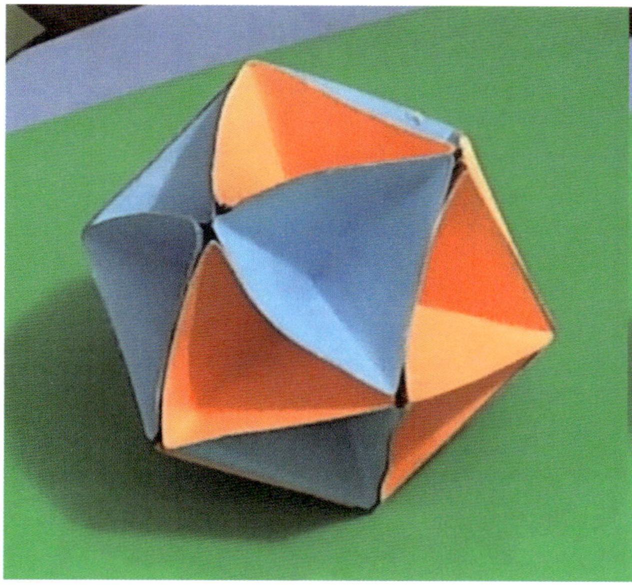

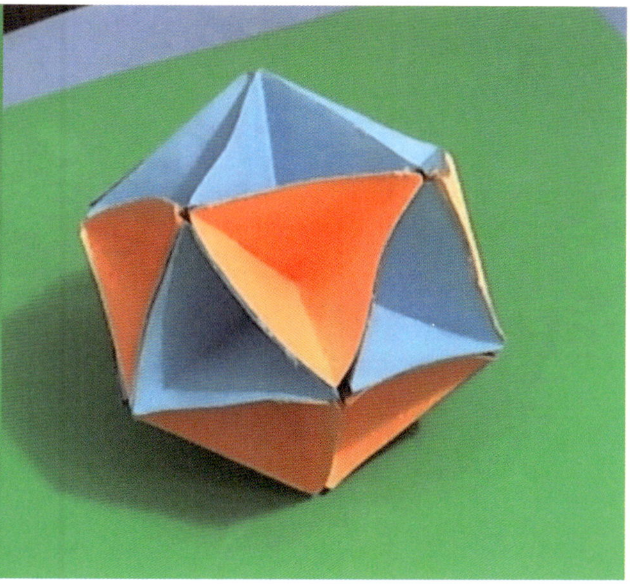

Icosidodecahedron Interior

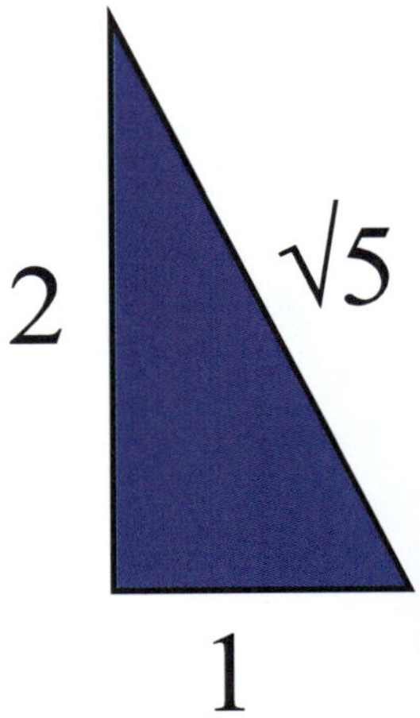

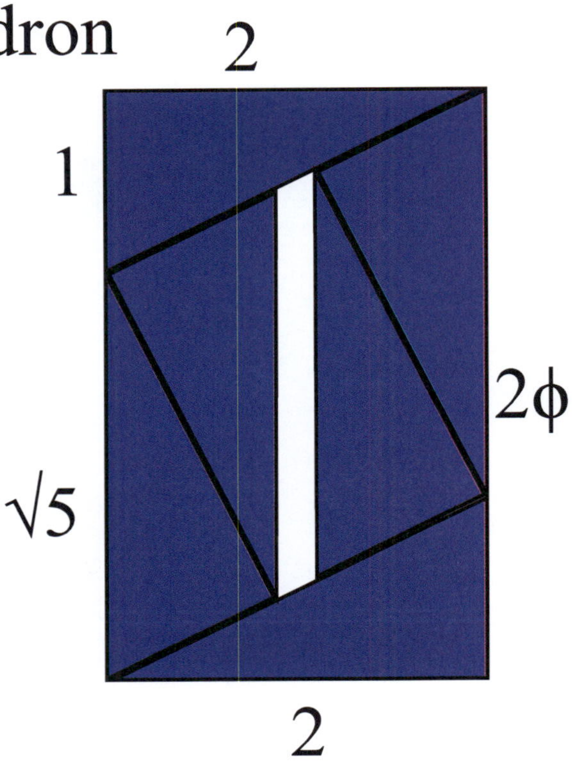

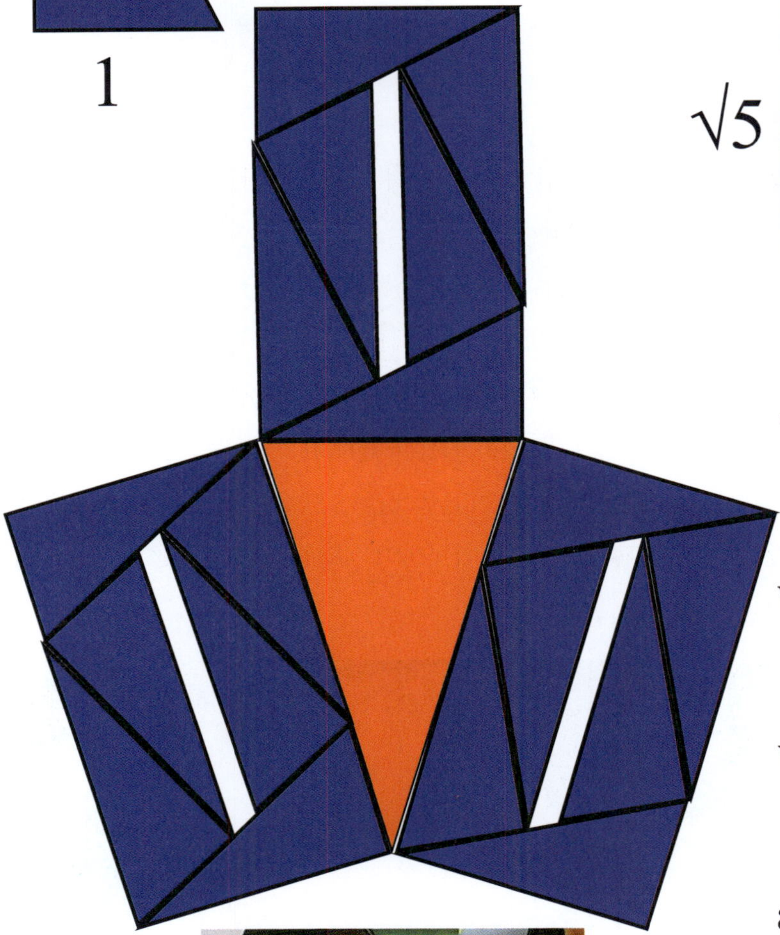

To Create the Interior of the Icosidodecahedron we resort to the Golden Rectangle again. We use this rectangle whose sides are in the Golden Ratio to create a very special triangle called the Golden Triangle. The sides and the base are also in the <u>Golden Ratio</u>.

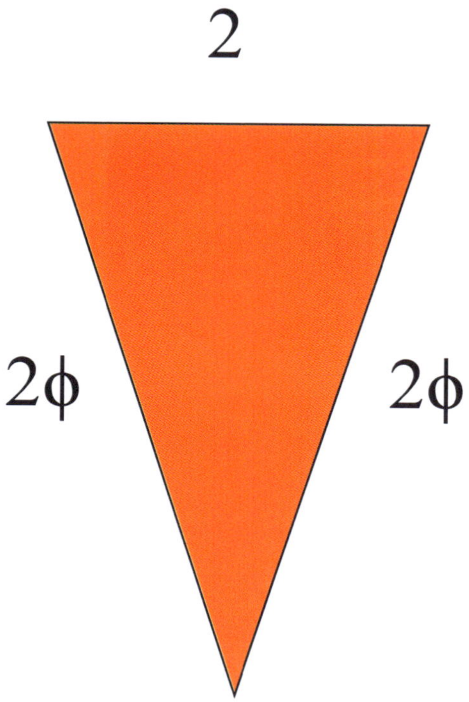

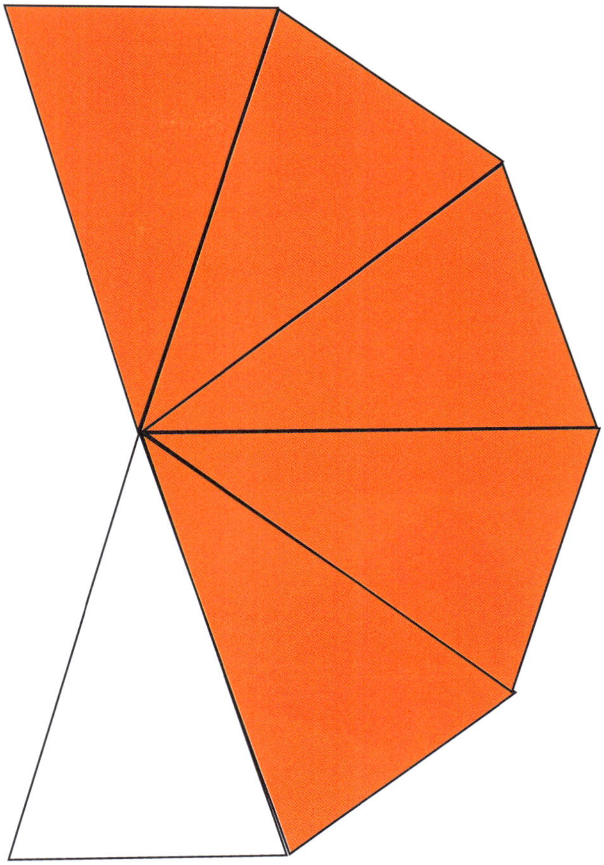

We array these triangles in groups of four and six because the Icosidodecahedron has faces of triangles and pentagons. Ten of these triangles arrayed in a circle form a decagon, the 10 sided polygon.

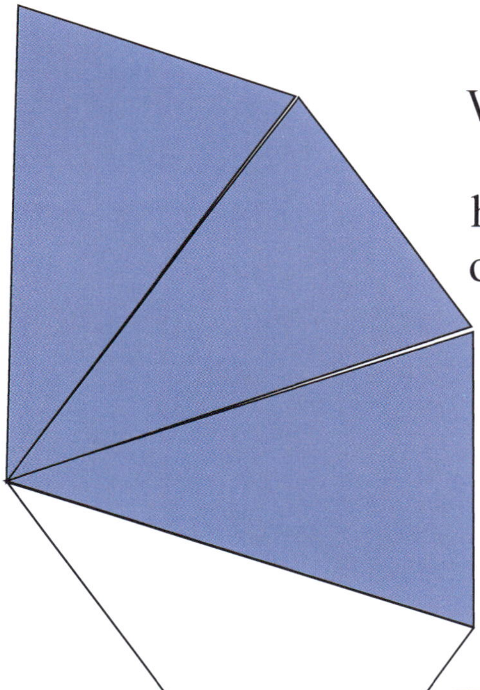

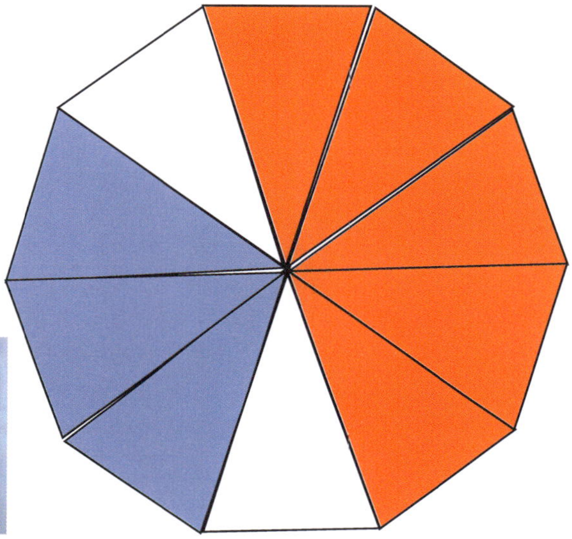

Rumor has it that the Icosidodecahedron was Leonardo Da Vinci's favorite shape. A friend of his did wood carvings of it and Leonardo did sketches of those. It's a very versatile structure. It is divisible along six axis, each of which reveals a decagon. The Golden Ratio is integrated into the interior triangles and in the Pentagonal Face. But perhaps its greatest asset is its beauty. It is wonderful to build and just to look at. It is truly one of mathematics' great treasures.

Cube-Octahedron Interior

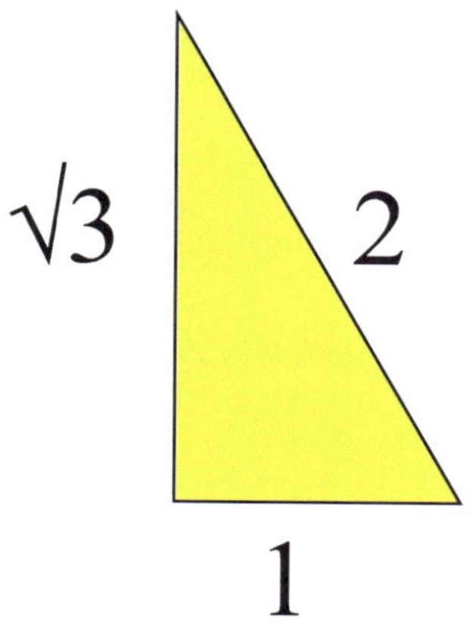

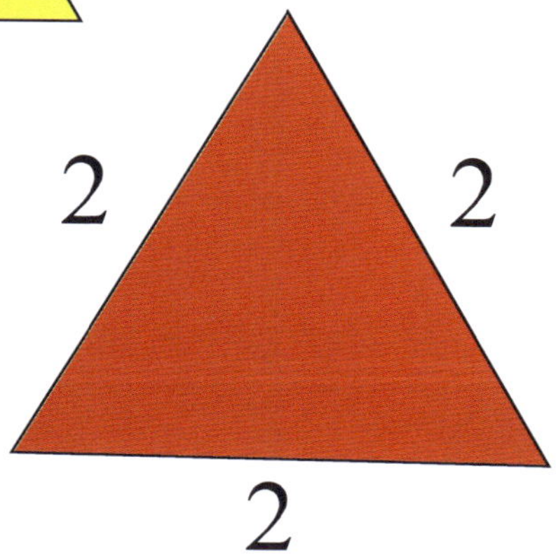

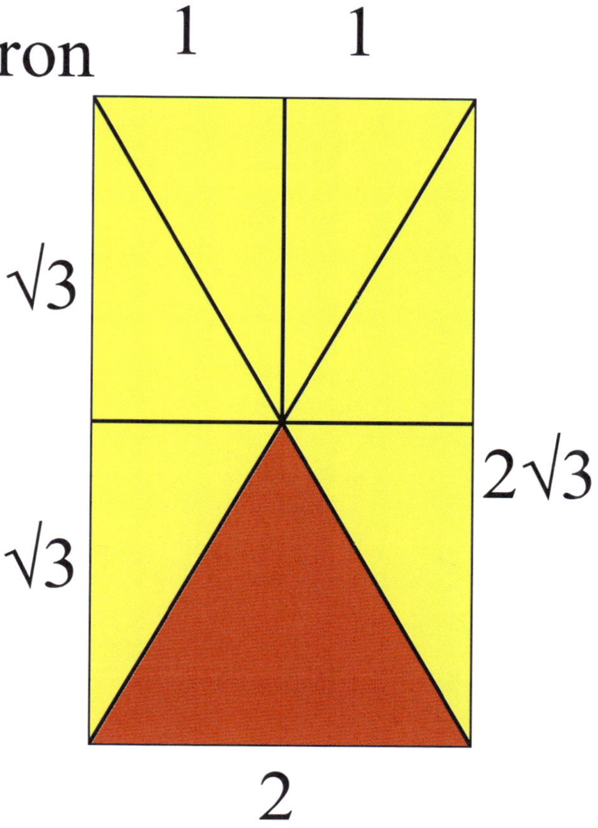

The Interior of the Cube-octahedron comes directly from the hexagon. The interior pyramids are the same shapes we built by folding the Hexagon and tucking triangles under.

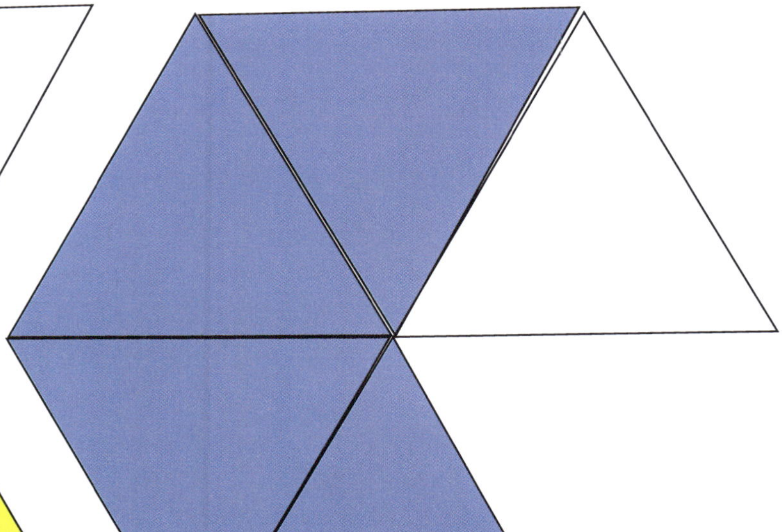

We array equilateral triangles around a common center. To construct the eight triangular pyramids we array four; three for the sides and one for the flap. For the pyramids that have four sides we array five triangles; four for the sides and one for the flap. We put them together to form the Cube-Octahedron, which Buckminster Fuller called Vector Equilibrium. He called it this because every edge and every internal strut are the same length. This has the great advantage of standardizing all building materials. It also has strength advantages; every linear connection is the same length, able to carry the same weight and stand up to the same stress.

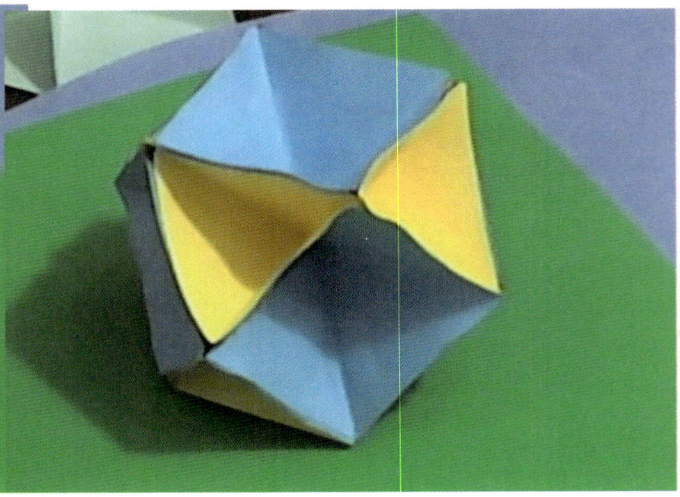

Inphinity

$1/(\phi+1)$
$1/\phi$
1
ϕ
$\phi+1$
$2\phi+1$
$3\phi+2$
$5\phi+3$
$8\phi+5$
$13\phi+8$
$21\phi+13$
$34\phi+21$
$55\phi+34$
$89\phi+55$

The Golden Ratio, Phi, Φ

Divide a line so that the ratio of the longer part to the shorter part is the same as the ratio of the whole line to the longer part. The Extreme and Mean Ratio.

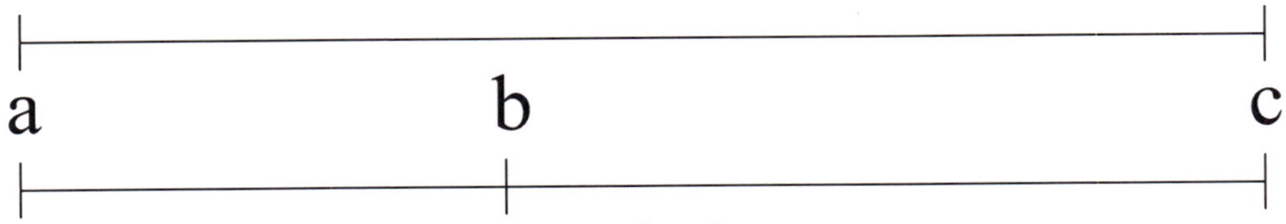

In Algebra.

$bc/ab = ac/bc = \Phi$ or $(\sqrt{5} + 1)/2 \approx 1.61.. \approx 2/(\sqrt{5} - 1)$

$ab/bc = bc/ac = 1/\Phi$ or $(\sqrt{5} - 1)/2 \approx 0.61.. \approx 2/(\sqrt{5} + 1)$

Phi first showed up in a unique way in our project when we noticed that one of our triangles, the 1,2,√5 had the three numbers of the formula for Phi, (√5 + 1)/2. We started playing around and we soon came up with an orientation that would give us **a** Golden Rectangle, but not **the** Golden Rectangle.

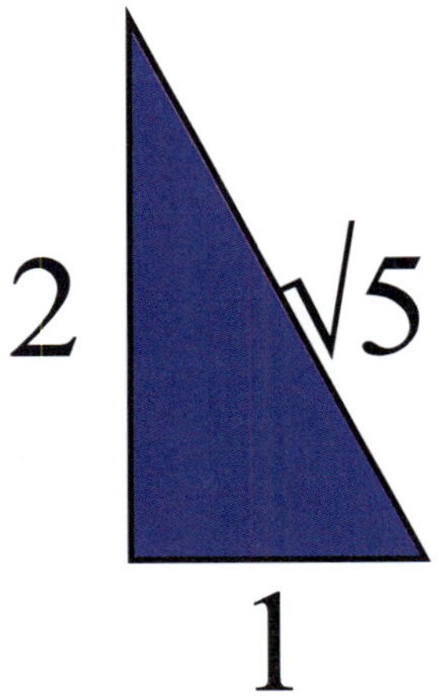

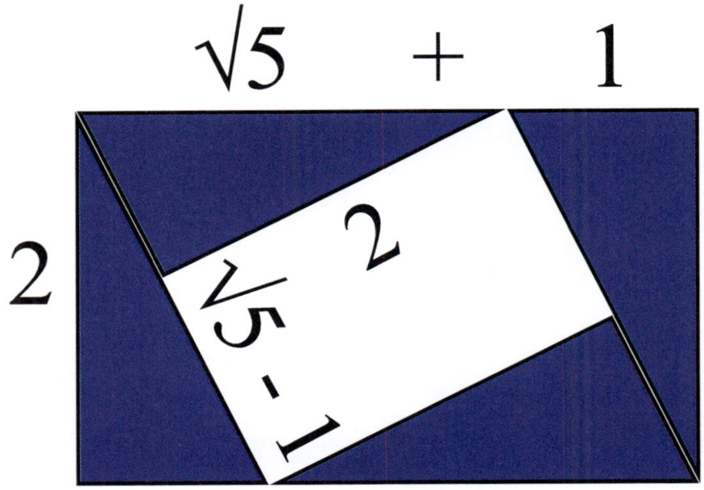

The ratio was correct but the sides were 2 and 2 Phi. What we really wanted was a rectangle that had sides of 1 and Phi. But before we started searching we were surprised to find that the rectangle in the middle was also a Golden Rectangle. The configuration was correct, but we needed to get the right length on the triangles.

Take a 1x1 Square

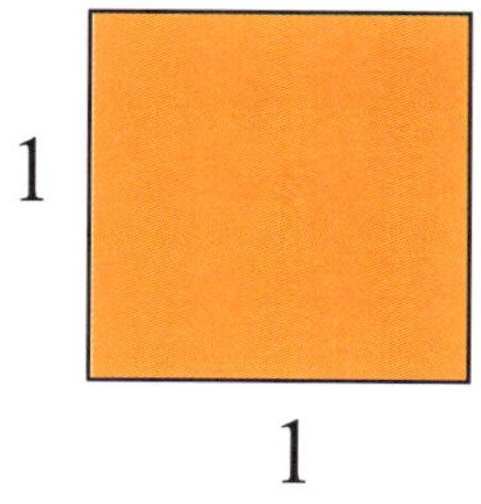

Slice it in two parts.

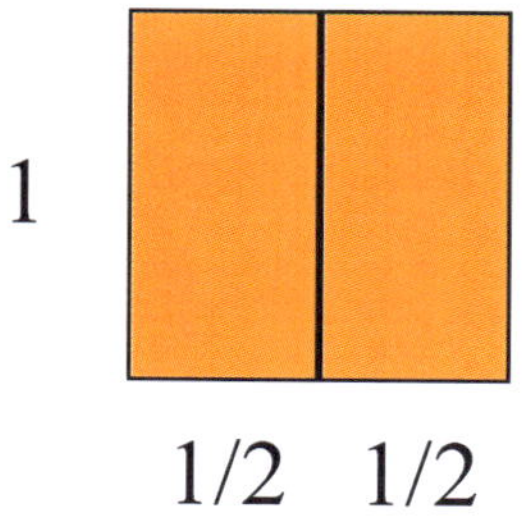

Cut these on the diagonal.

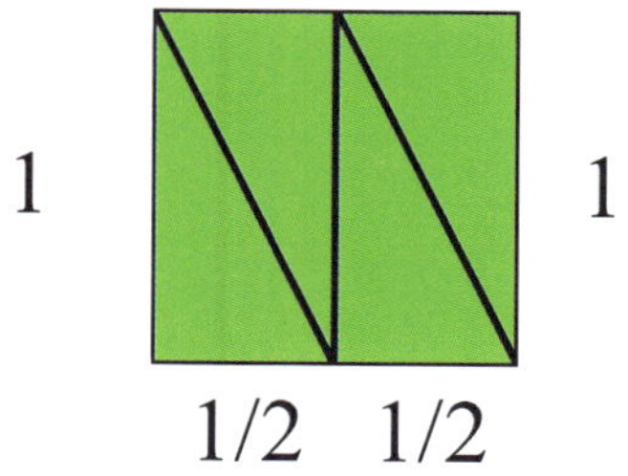

We get our triangle.

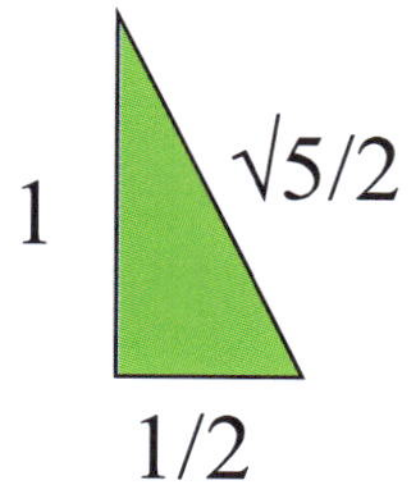

We assemble a rectangle. The longer side is 1 and the shorter side is √5/2 + 1/2. (√5 + 1)/2 is the definition of Phi, ϕ.

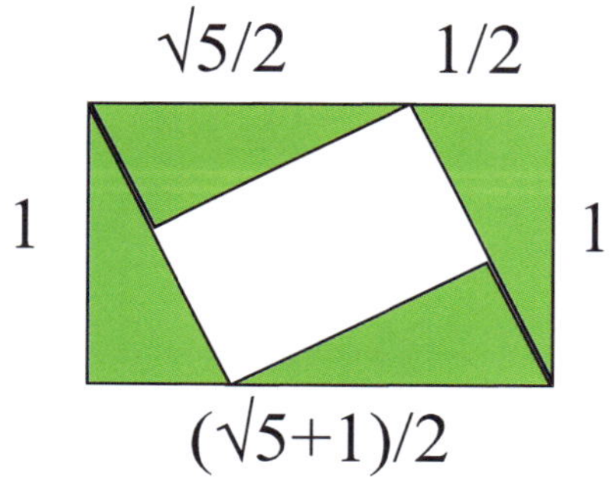

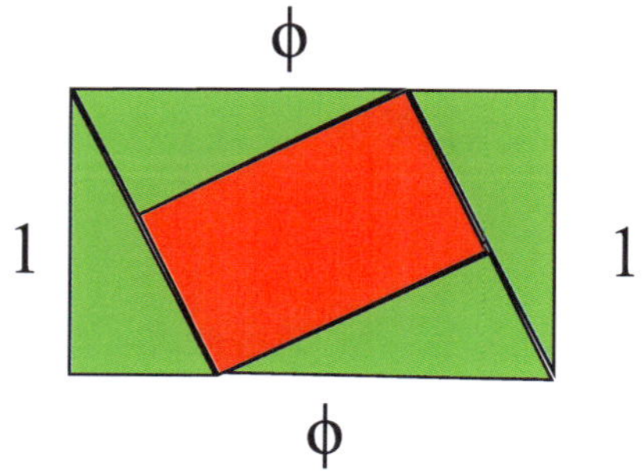

The center rectangle is amazing as well. It is also a Golden Rectangle. Its dimensions, 1x(√5-1)/2, or 1x(1/ϕ).

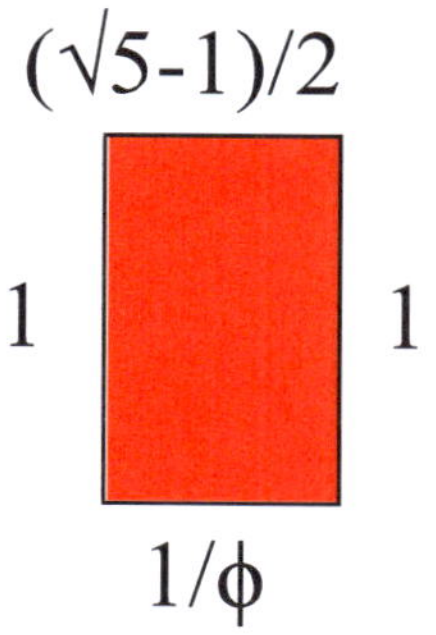

This puts these two rectangles in a unique progression. (1/ϕ), 1, ϕ They are succeeding multiples of ϕ.

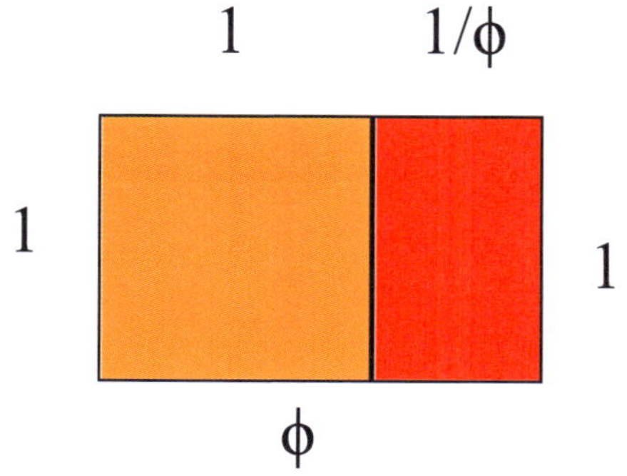

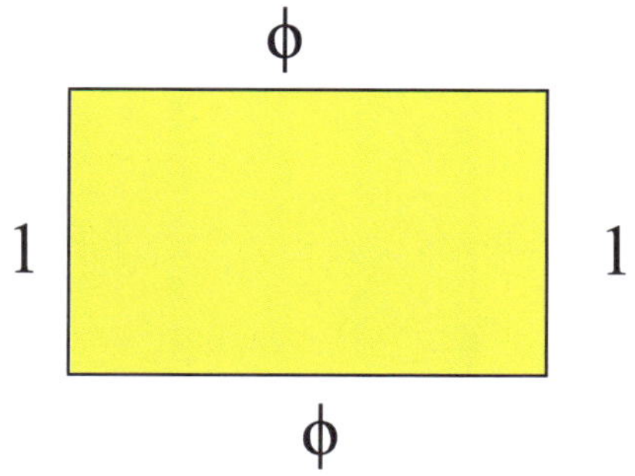

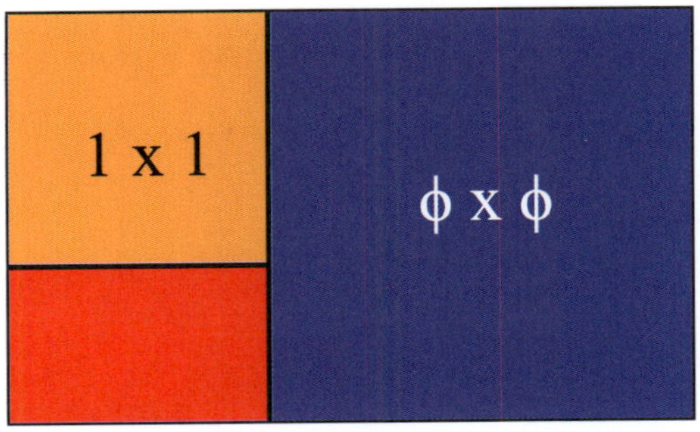

This process can be continued. You can place a square to the long side of the Golden Rectangle. This will create the next Golden Rectangle.

Or you can slice up the square into four triangles and place them around. Each new square is a multiple of ϕ.

This is like the Fibonacci Series, except instead of approaching the Golden Ratio as a Limit, these are exact representations of the Golden Ratio.

This Series is an odd combination of a linear and a geometric progression. You can multiply by ϕ (1.61828..) each time or simply add the two previous.

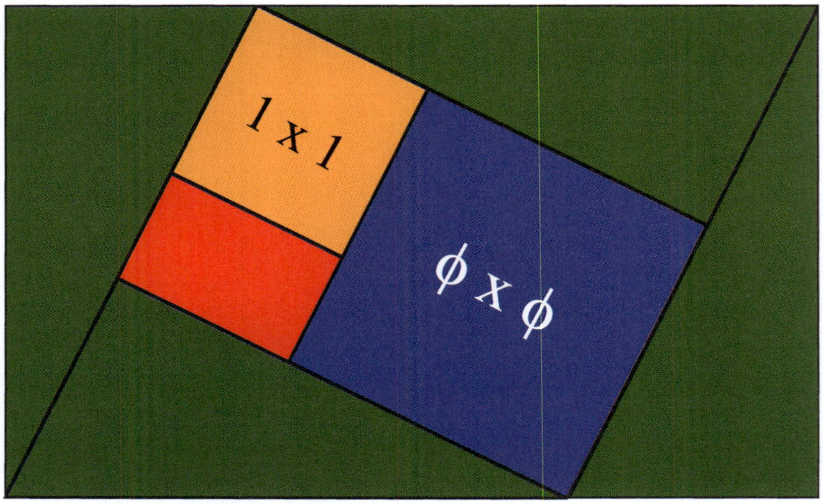

These are algebraic representations of the successive sides. Here you can see clearly that they are succeeding multiples of ϕ. You can also see that each succeeding number is the sum of the two previous. This progression continues into infinity, or Inphinity.

The Fibonacci Series, which has made this process well known, begins with whole numbers. 1, 1, 2, 3, 5, 8, 13, 21, 34, 55, 89, 144. You add the two previous numbers together and you get the next number in the series. The Ratio of the Successive numbers gets closer and closer to ϕ, 1.61828..

$$\phi^{-1} = 1/\phi$$

$$\phi^{0} = 1$$

$$\phi^{1} = \phi$$

$$\phi^{2} = \phi + 1$$

$$\phi^{3} = 2\phi+1$$

$$\phi^{4} = 3\phi+2$$

$$\phi^{5} = 5\phi+3$$

While we use 1 Inch as the size of our Unit Square in this book, any size will do. We recommend three other sizes, the Decimeter, the Foot, and the Meter. This gives a good range and does not tie one to a particular measurement system.

Fibonacci Sequence

0
1
1
2
3
5
8
13
21
34
55
89
144
233

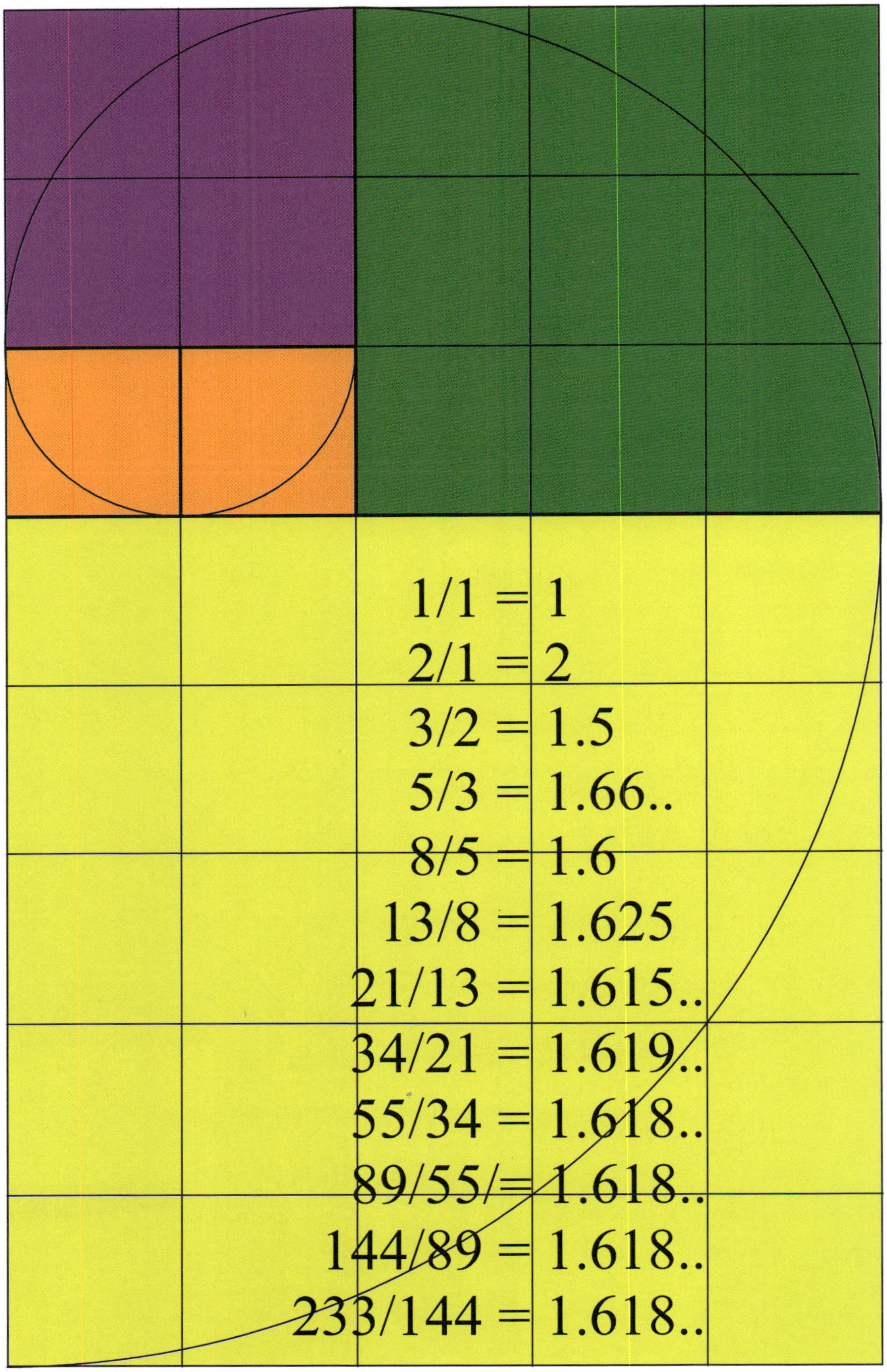

Each rectangle gets closer to Phi.

The Phi Sequence

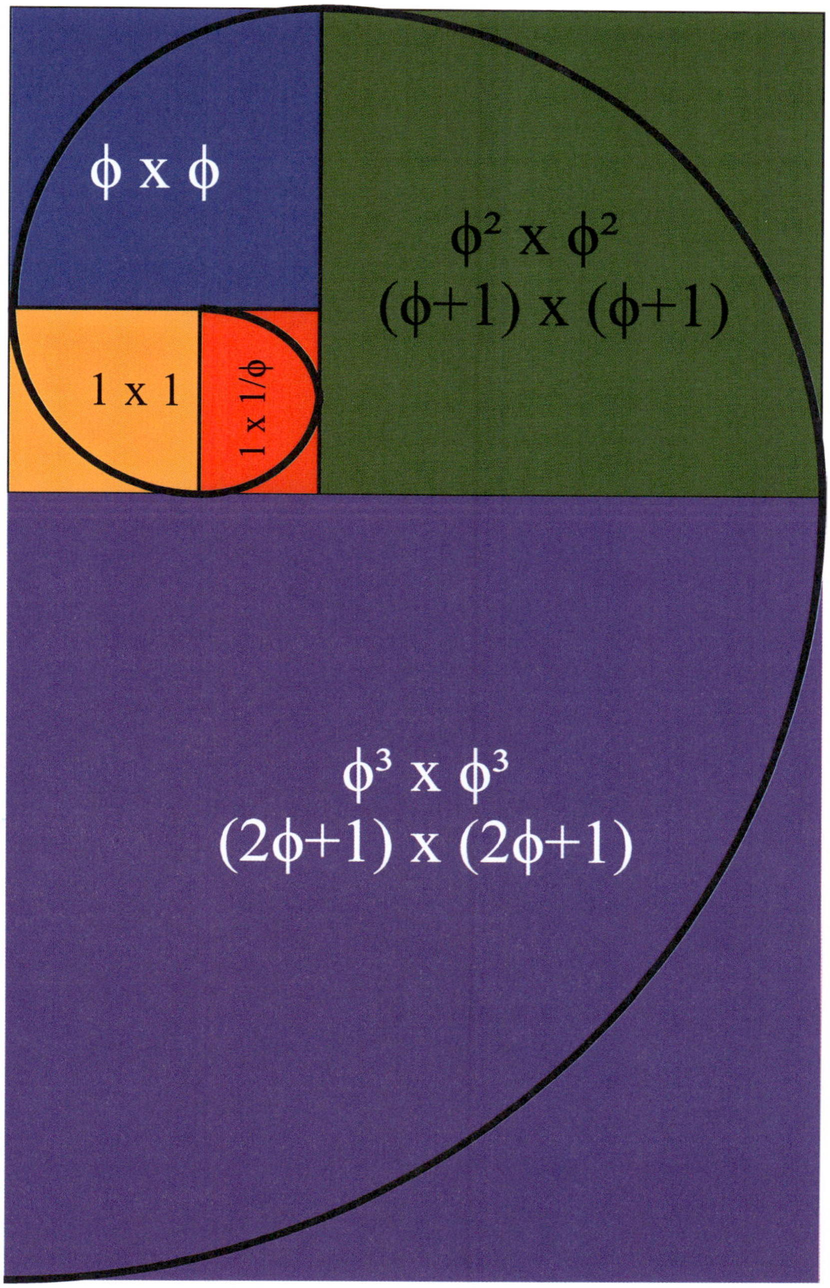

$1/\phi$	0.6182..
1	1
ϕ	1.6180..
$\phi+1$	2.6180..
$2\phi+1$	4.2360..
$3\phi+2$	6.8540..
$5\phi+3$	11.091..
$8\phi+5$	17.944..
$13\phi+8$	29.034..
$21\phi+13$	46.978..
$34\phi+21$	76.012..
$55\phi+34$	122.99..
$89\phi+55$	199.11..

This sequence doesn't just approach Phi, each new rectangle is exactly in the Golden Ratio. You get the next side by multiplying by Phi or adding the previous two sides together.

The Golden Spiral

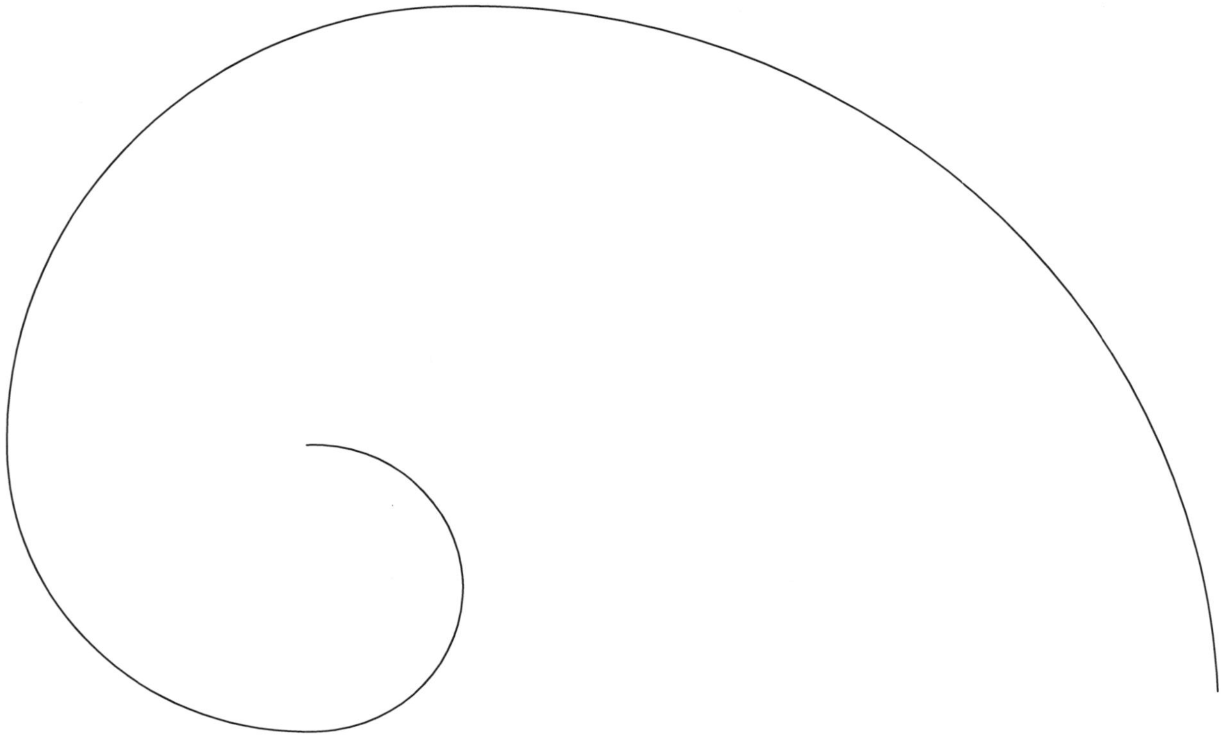

The Golden Spiral is a special case of the Logarithmic Spiral, or as it is known, the Equiangular Spiral. The Golden Spiral is a Logarithmic Spiral whose growth rate is Phi for each Quarter Turn. This is the shape in which the transcendental number ϕ meets the transcendental number e.

e is nature's growth constant. It is also the base of the natural logarithms. e to the x is known as the exponential function. By allowing the Logarithmic Spiral to combine its many characteristics with the Golden Ratio we create an entity that has many manifestations in nature. The Nautilus Shell, Spiral Storms, and Galaxies are all examples of this combination.

The Golden Triangle

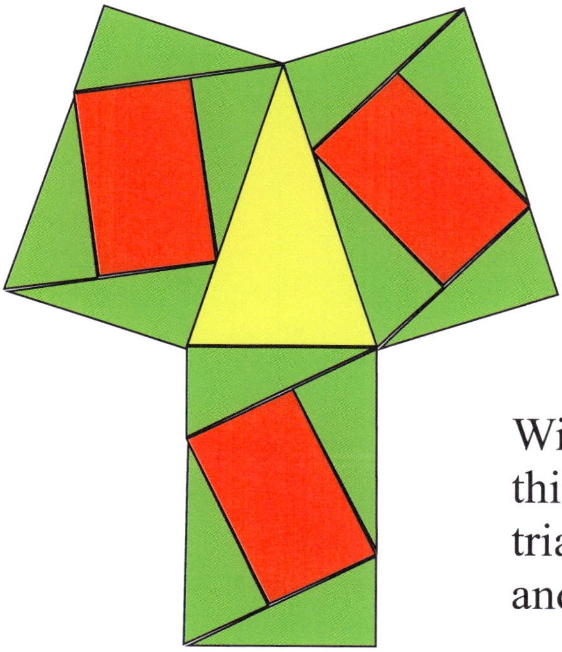

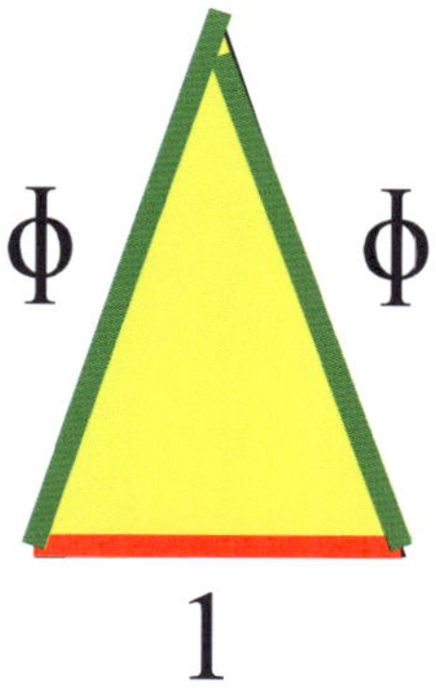

With a base of one, the legs of this isosceles triangle are ϕ. These triangles appear in the decagon and on the arms of the pentangle.

Kepler's Triangle

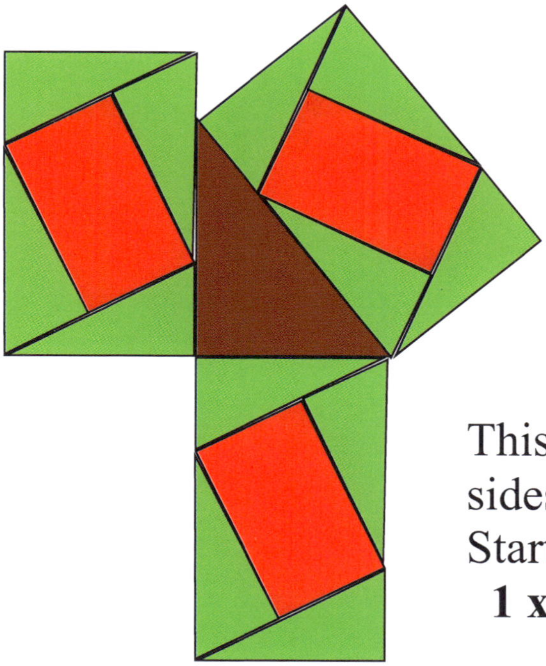

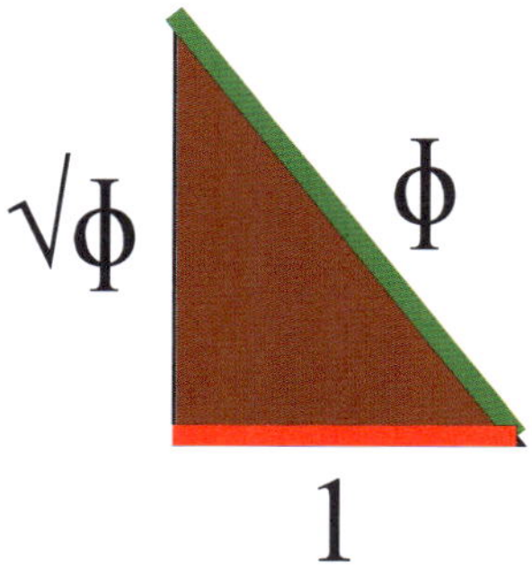

This is the only right triangle whose sides are in a geometric progression. Starting with 1 the sides grow by √ϕ.

$\mathbf{1} \times \sqrt{\phi} = \sqrt{\phi} \qquad \sqrt{\phi} \times \sqrt{\phi} = \phi$

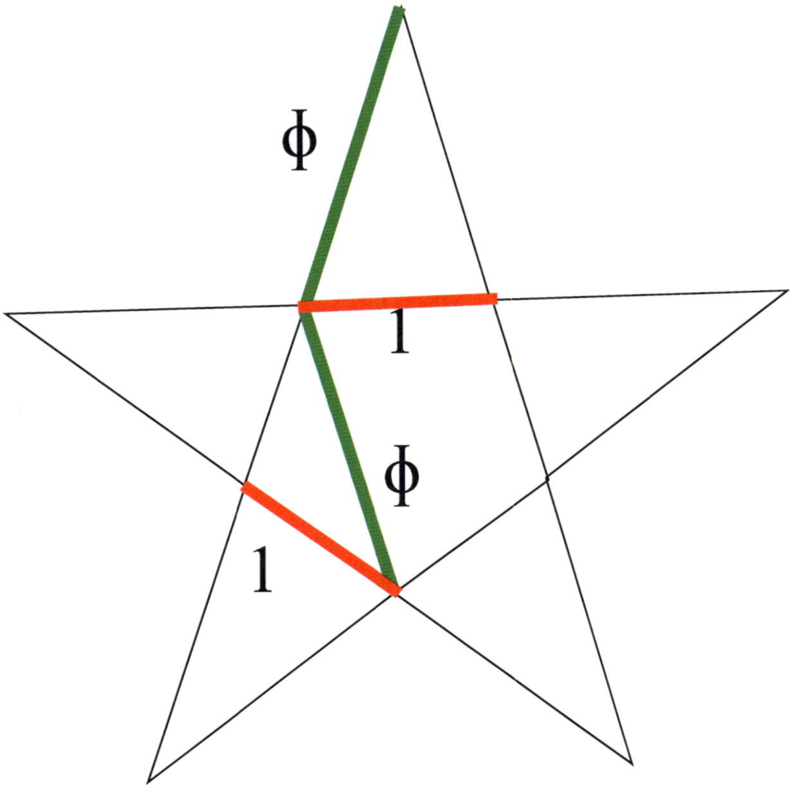

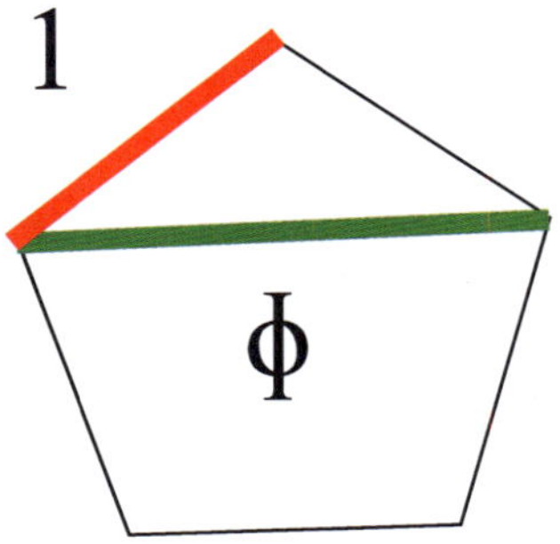

The Decagon is 10 Golden Triangles arrayed around a single point.

$2\phi\pi$

The Circle around the Decagon brings together ϕ and π.

Decagon Edge - 1
Circle Radius - ϕ
Circle Diameter - 2ϕ
Circle Circumference - $2\phi\pi$

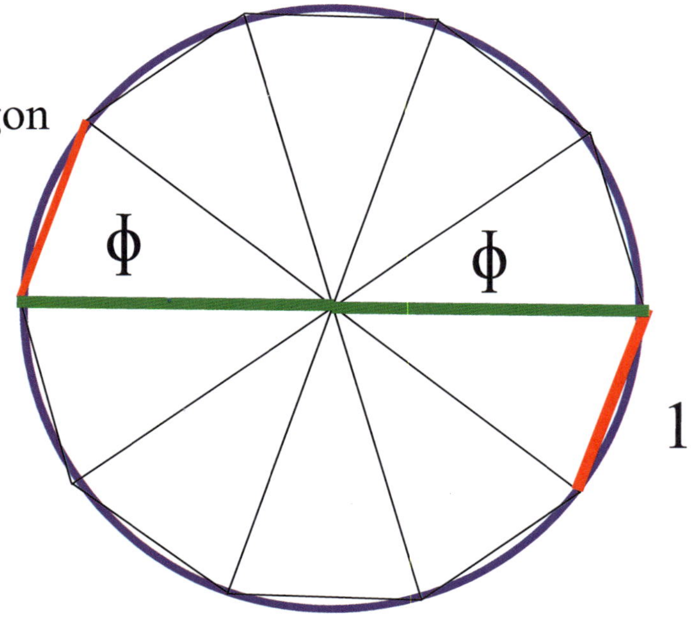

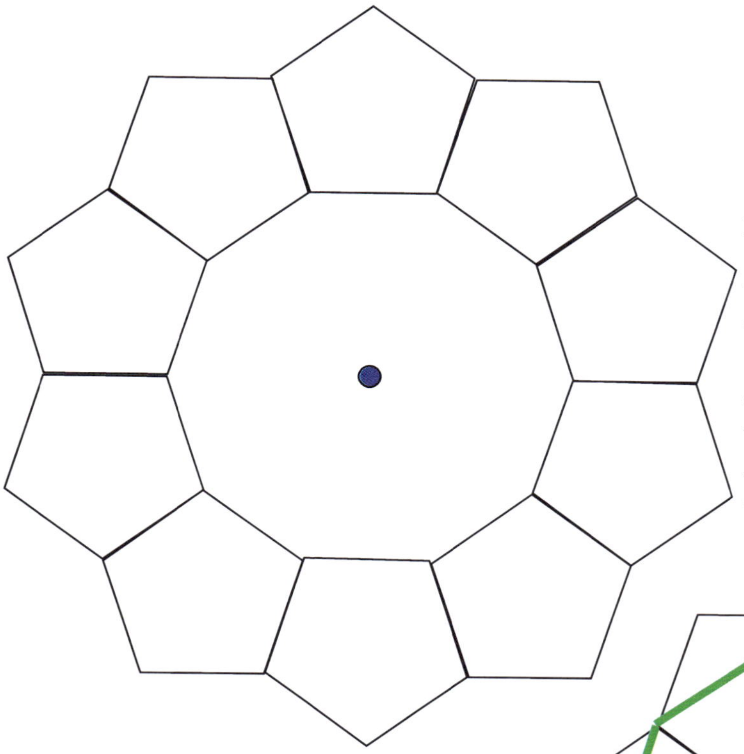

Ten Pentagons joined together form an internal decagon. We mark these lines red. We make another decagon by connecting two of the corners in the pentagons with a green line. If the red line is 1 the green line is Phi, ϕ.

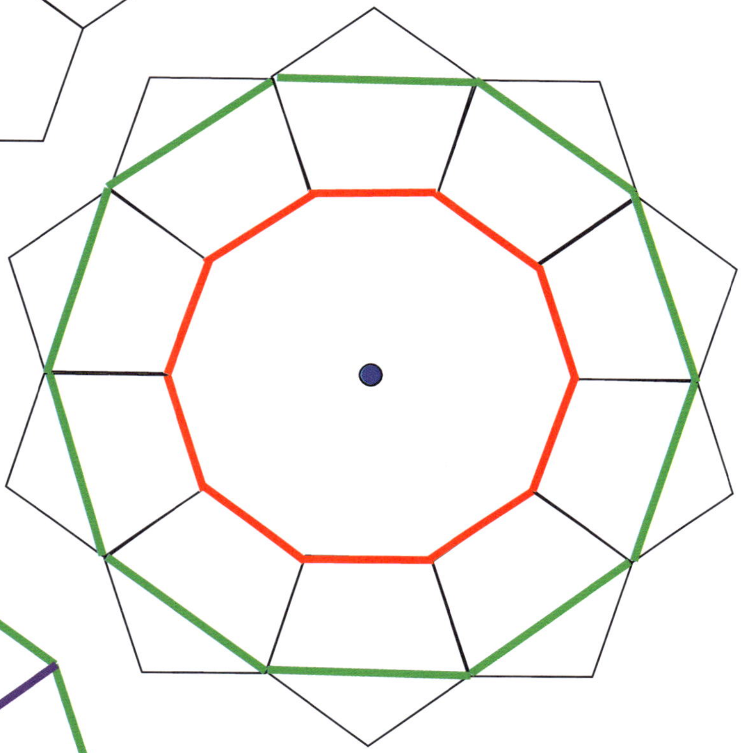

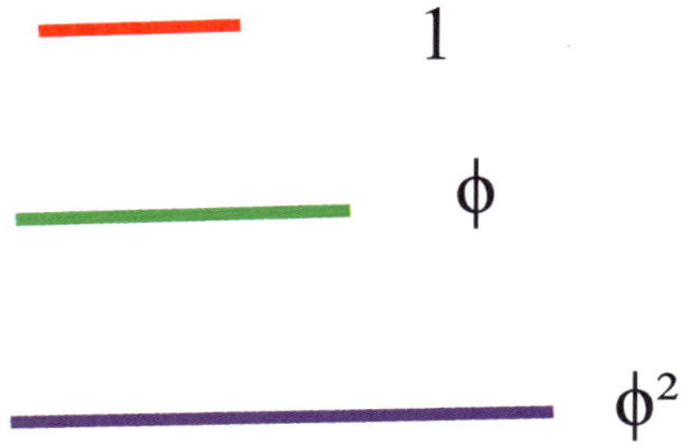

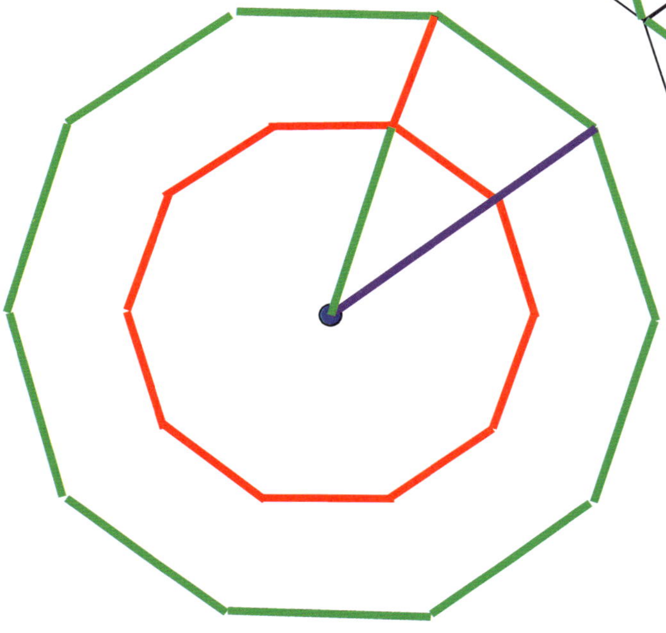

We have created a decagon with sides of one, another decagon with sides of ϕ, two interlocking large pentagons with sides of $\phi + 2$. The radii of the various shapes are ϕ, and ϕ^2.

This structure is made entirely of Golden Triangles and Golden Rectangles. The blue and purple center is a Decagon composed of ten Golden Triangles. Around the central Decagon are ten Golden Rectangles with a Golden Triangle between each one. The twenty outer edges of these shapes form an Icosagon of twenty equal edges. Surrounding the Icosagon are ten Golden Rectangles and ten rhombi made of 3 Golden Triangles each.

Printed in Dunstable, United Kingdom

91222491R00060